Mathematical Foundations for Electromagnetic Theory

IEEE PRESS Series on Electromagnetic Waves

The IEEE PRESS Series on Electromagnetic Waves consists of new titles as well as reprints and revisions of recognized classics that maintain long-term archival significance in electromagnetic waves and applications.

Series Books Published

Collin, R. E., *Field Theory of Guided Waves*, 2nd ed., 1991
Dudley, D. G., *Mathematical Foundations for Electromagnetic Theory*, 1994
Elliott, R. S., *Electromagnetics: History, Theory, and Applications*, 1993
Felsen, L. B., and Marcuvitz, N., *Radiation and Scattering of Waves*, 1994
Harrington, R. F., *Field Computation by Moment Methods*, 1993
Tai, C. T., *Dyadic Green Functions in Electromagnetic Theory*, 2nd ed., 1993
Tai, C. T., *Generalized Vector and Dyadic Analysis: Applied Mathematics in Field Theory*, 1991

Mathematical Foundations for Electromagnetic Theory

Donald G. Dudley
University of Arizona, Tucson

IEEE PRESS Series on
Electromagnetic Waves
Donald G. Dudley, Series Editor

IEEE
PRESS

IEEE Antennas and Propagation Society, *Sponsor*

The Institute of Electrical and Electronics Engineers, Inc., New York

IEEE PRESS
445 Hoes Lane, PO Box 1331
Piscataway, NJ 08855-1331

This book may be purchased at a discount from the publisher when ordered in bulk quantities. For more information contact:

IEEE PRESS Marketing
Attn: Special Sales
P.O. Box 1331
445 Hoes Lane
Piscataway, NJ 08855-1331

Printed in the United States of America
10 9 8 7 6 5 4 3 2 1

ISBN 0-7803-1022-5

IEEE Order Number: PC03715

Library of Congress Cataloging-in-Publication Data
Dudley, Donald G.
Mathematical foundations for electromagnetic theory / Donald G. Dudley.
p. cm. — (IEEE electromagnetic waves series)
"IEEE Antennas and Propagation Society, sponsor."
Includes bibliographical references and index.
ISBN 0-7803-1022-5
1. Electromagnetic theory—Mathematics. I. IEEE Antennas and Propagation Society." II. Title. III. Series.
QC670.D79 1994 94-3159
530.1′41′0151—dc20 CIP

Dedicated to Professor Robert S. Elliott

Contents

Preface

This book is written for the serious student of electromagnetic theory. It is a principal product of my experience over the past 25 years interacting with graduate students in electromagnetics and applied mathematics at the University of Arizona.

A large volume of literature has appeared since the latter days of World War II, written by researchers expanding the basic principles of electromagnetic theory and applying the electromagnetic model to many important practical problems. In spite of widespread and continuing interest in electromagnetics, the underlying mathematical principles used freely throughout graduate electromagnetic texts have not been systematically presented in the texts as preambles. This is in contrast to the situation regarding undergraduate electromagnetic texts, most of which contain preliminary treatments of fundamental applied mathematical principles, such as vector analysis, complex arithmetic, and phasors. It is my belief that there should be a graduate electromagnetic theory text with linear spaces, Green's functions, and spectral expansions as mathematical cornerstones. Such a text should allow the reader access to the mathematics and the electromagnetic applications without the necessity for consulting a wide range of mathematical books written at a variety of levels. This book is an effort to bring the power of the mathematics to bear on electromagnetic problems in a single text.

Since the mastery of the foundations for electromagnetics provided in this book can involve a considerable investment of time, I should like to indicate some of the potential rewards. When the student first begins a study of electromagnetic theory at the graduate level, he/she is confronted with a large array of series expansions and transforms with which to reduce the differential equations and boundary conditions in a wide variety of canonical problems in Cartesian, cylindrical, and spherical coordinates. Often, it seems to the student that experience is the only way to determine specifically which expansions or transforms to use in a given problem. In addition, convergence properties seem quite mysterious. These issues can be approached on a firm mathematical base through the foundations provided in this book. Indeed, the reader will find that different differential operators with their associated boundary conditions lead to specific expansions and transforms that are "natural" in a concrete mathematical sense for the problem being considered. My experience with graduate students has been that mastery of the foundations allows them to appreciate why certain expansions and transforms are used in the study of canonical problems. Then, what is potentially more important, the foundations allow them to begin the more difficult task of formulating and solving problems on their own.

I first became interested in Green's functions and spectral representations during my graduate studies at UCLA in the 1960s. I was particularly influenced by the treatment of the spectral representations of the delta function by Bernard Friedman [1], whose book at that time formed the cornerstone of the Applied Mathematics Program at UCLA in the College of Engineering. Subsequently, examples of spectral representations began appearing in texts on electromagnetic theory, such as [2]–[4], and, more recently, [5]. However, no text specifically devoted to Green's functions and spectral expansions and their application to electromagnetic problems has been forthcoming.

The material in this book forms a two-semester sequence for graduate students at the University of Arizona. The first three chapters contain the mathematical foundations, and are covered in a course offered every year to electrical engineering and applied mathematics graduate students with a wide range of interests. Indeed, the first three chapters in this book could be studied by applied mathematicians, physicists, and engineers with no particular interest in the electromagnetic applications. The fourth and fifth chapters are concerned with the electromagnetics, and are covered in a course on advanced electromagnetic theory, offered biennially. In this book, I have presumed that the reader has a working knowledge of complex variables. In addition, in the last two chapters, I have assumed that the reader has studied an introductory treatment of electromagnetics at the graduate level, as can be found, for example, in the texts by Harrington

[6], Ishimaru [7], or Balanis [8]. I have therefore felt no necessity to include a chapter on Maxwell's equations or a chapter on analytic function theory, presupposing reader familiarity.

Chapter 1 is an introduction to modern linear analysis. It begins with the notion of a linear space. Structure is added by the introduction of the inner product and the norm. With the addition of suitable convergence criteria, the space becomes a Hilbert space. Included in the discussion of Hilbert space are the concepts of best approximation and projection. The chapter concludes with a discussion of operators in Hilbert space. Emphasis is placed on the matrix representation of operations, a concept that leads naturally to the Method of Moments, one of the most popular techniques for the numerical solution to integral equations occurring in electromagnetic boundary value problems.

Chapter 2 covers Green's functions for linear, ordinary, differential operators of second order. The chapter begins with a discussion of the delta function. The Sturm–Liouville operator is introduced and discussed for three cases, which we title SLP1, SLP2, and SLP3. A clear distinction is made between self-adjoint and nonself-adjoint operators. In addition, the concepts of limit point and limit circle cases are introduced and explored through examples applicable to electromagnetic problems.

Chapter 3 introduces the spectral representation of the delta function. The theory is applied by example to various operators and boundary conditions. Included are important representations associated with the limit point and limit circle cases introduced in the previous chapter. A wide variety of spectral representations are presented in a form suitable for use in solving electromagnetic boundary value problems in multiple dimensions. These representations are augmented by further examples in the Problems.

Chapter 4 contains a discussion of fundamental electromagnetic sources represented by delta functions. The sources are analyzed using spectral representations and Green's functions in Cartesian, cylindrical, and spherical conditions. A variety of useful alternative representations emerge. Included are sheet sources, line sources, ring sources, shell sources, and point sources.

In Chapter 5, the ideas developed in the previous chapters are applied to a sample of electromagnetic boundary value problems. No attempt is made to produce an exhaustive collection. Rather, the purpose of the chapter is to demonstrate the power of the structure developed in the first three chapters. Static problems included involve the rectangular box and rectangular cylinder. Dynamic problems include propagation in a parallel plate waveguide, scattering by an iris obstacle in a parallel plate waveguide, aperture diffraction, and scattering by a conducting cylinder. Emphasis has been placed on the power of alternative representations by including useful alternatives in the examples on the parallel plate waveguide and scattering from a conducting circular cylinder.

My graduate students over the past 25 years have had a major influence on this book. All have contributed through classroom and individual discussions. Many too numerous to mention have made suggestions and corrections in early drafts. Specifically, I should like to acknowledge some special help. In the early 1980s, K. A. Nabulsi and Amal Nabulsi painstakingly typed a portion of my handwritten class notes. These typed notes were produced before the advent of modern computational word processors, and formed the basis for my subsequent writing of Chapters 1–3 of this book. Dr. Nabulsi, now a Professor in Saudi Arabia, sent me a gifted student, Muntasir Sheikh, for doctoral training. Mr. Sheikh has critically read the entire book manuscript and offered suggestions and corrections. In addition, Charles Trantanella, Michael Pasik, and Jacob Adopley have carefully read portions of the manuscript.

In the mid-1970s, I had the good fortune to be a part of the creation of the now greatly successful Program in Applied Mathematics at the University of Arizona. W. A. Johnson was my first student to graduate through the program. Because of him, I became acquainted with three professors in the Department of Mathematics, C. L. DeVito, W. M. Greenlee, and W. G. Faris. These four mathematicians have had a lasting influence on the way I have come to consider many of the mathematical issues involved in electromagnetic theory.

Among my colleagues, there are several who have had a marked influence on this book. R. E. Kleinman, University of Delaware, has consistently encouraged me to pursue my mathematical interests applied to electromagnetic theory. L. B. Felsen, Polytechnic University, has influenced me in many ways, scientifically and personally. In addition, his comments concerning modern research applications led me to some important additions in Chapter 5. K. J. Langenberg, University of Kassel, has read in detail the first three chapters and offered important advice and criticism. R. W. Ziolkowski, University of Arizona, has taught a course using the material contained in Chapters 1–3 and offered many suggestions and corrections. I. Stakgold, University of Delaware, made me aware of the recent mathematical literature on limit point and limit circle problems.

Many reviewers, anonymous and known, have made comments that have led me to make changes and additions. I would particularly like to mention Ehud Heyman, Tel Aviv University, whose comments concerning alternative representations led me to strengthen this material in Chapter 5. I would also like to thank Dudley Kay and the staff at IEEE Press whose competence and diligence have been instrumental in the production phase of this book project.

With Chalmers M. Butler, Clemson University, a distinguished educator and cherished friend, I have had the good fortune to have a 20-year running discussion concerning methods of teaching electromagnetics to graduate students. Part of the fun has been that we have not always agreed. However, one issue upon which there has been no disagreement is the importance of presenting electromagnetics to students in a structurally organized manner, stressing the common links between wide ranges of problems. I have drawn strength, satisfaction, and pleasure from our association.

My family has always seemed to understand my many interests, and this book has been a major one for more years than I should like to recall. It is with love and affection that I acknowledge my wife, Marjorie A. Dudley; my children, Donald L. Dudley and Susan D. Benson; and the memory of my former wife, Marjorie M. Dudley. Love truly does "make the world go 'round."

Finally, it is with gratitude that I dedicate this book to my teacher, mentor, and friend, Robert S. Elliott, University of California at Los Angeles, a consummate scholar without whom none of this would have occurred.

REFERENCES

[1] Friedman, B. (1956), *Principles and Techniques of Applied Mathematics.* New York: Wiley.

[2] Jackson, J. D. (1962), *Classical Electrodynamics.* New York: Wiley.

[3] Collin, R. E. (1960, 2nd edition 1991), *Field Theory of Guided Waves.* New York: IEEE Press.

[4] Felsen, L. B., and N. Marcuvitz (1973), *Radiation and Scattering of Waves.* Englewood Cliffs, NJ: Prentice-Hall.

[5] Ishimaru, A. (1991), *Electromagnetic Wave Propagation, Radiation, and Scattering.* Englewood Cliffs, NJ: Prentice-Hall.

[6] Harrington, R. F. (1961, reissued 1987), *Time-Harmonic Electromagnetic Fields.* New York: McGraw-Hill.

[7] Op. cit. Ishimaru.

[8] Balanis, C. A. (1989), *Advanced Engineering Electromagnetics.* New York: Wiley.

1

Linear Analysis

1.1 INTRODUCTION

Fundamental to the study of many of the differential equations describing physical processes in applied physics and engineering is *linear analysis.* Linear analysis can be elegantly and logically placed in a mathematical structure called a *linear space*.

We begin this chapter with the definition of a linear space. We then begin to add structure to the linear space by introducing the concepts of inner product and norm. Our study leads us to Hilbert space and, finally, to linear operators within Hilbert space. The characteristics of these operators are basic to the ensuing development of the differential operators and differential equations found in electromagnetic theory.

Throughout this chapter, we shall be developing notions concerning vectors in a linear space. These ideas make use of both the real and complex number systems. A knowledge of the axioms and theorems governing real and complex numbers will be assumed in what follows. We shall use this information freely in the proofs involving vectors.

1.2 LINEAR SPACE

Let $a, b, c, \ldots$ be elements of a set $\mathcal{S}$. These elements are called *vectors*. Let $\alpha, \beta, \ldots$ be elements of the field of numbers $\mathbf{F}$. In particular, let $\mathbf{R}$ and

C be the field of real and complex numbers, respectively. The set $\mathcal{S}$ is a *linear space* if the following rules for addition and multiplication apply:

I. Rules for addition among vectors in $\mathcal{S}$:

a. $(a+b)+c = a+(b+c)$
b. There exists a zero vector $\mathbf{0}$ such that $a+\mathbf{0} = \mathbf{0}+a = a$.
c. For every $a \in \mathcal{S}$, there exists $-a \in \mathcal{S}$ such that $a+(-a) = (-a)+a = \mathbf{0}$.
d. $a+b = b+a$

II. Rules for multiplication of vectors in $\mathcal{S}$ by elements of **F**:

a. $\alpha(\beta a) = (\alpha\beta)a$
b. $1a = a$
c. $\alpha(a+b) = \alpha a + \alpha b$
d. $(\alpha+\beta)a = \alpha a + \beta a$

EXAMPLE 1.1 Consider *Euclidean space* $\mathbf{R}_n$. Define vectors a and b in $\mathbf{R}_n$ as follows:

$$a = (\alpha_1, \alpha_2, \ldots, \alpha_n) \tag{1.1}$$

$$b = (\beta_1, \beta_2, \ldots, \beta_n) \tag{1.2}$$

where α_k and β_k, the *components* of vectors a and b, are in **R**, $k = 1, 2, \ldots, n$. Define addition and multiplication as follows:

$$\begin{aligned} a+b &= (\alpha_1, \ldots, \alpha_n) + (\beta_1, \ldots, \beta_n) \\ &= (\alpha_1+\beta_1, \ldots, \alpha_n+\beta_n) \end{aligned} \tag{1.3}$$

$$\begin{aligned} \alpha a &= \alpha(\alpha_1, \ldots, \alpha_n) \\ &= (\alpha\alpha_1, \ldots, \alpha\alpha_n) \end{aligned} \tag{1.4}$$

where $\alpha \in \mathbf{R}$. If we assume prior establishment of rules for addition and multiplication in the field of real numbers, it is easy to show that $\mathbf{R}_n$ is a linear space. We must show that the rules in I and II are satisfied. For example, for addition rule d,

$$\begin{aligned} a+b &= (\alpha_1+\beta_1, \ldots, \alpha_n+\beta_n) \\ &= (\beta_1+\alpha_1, \ldots, \beta_n+\alpha_n) \\ &= b+a \end{aligned} \tag{1.5}$$

We leave the satisfaction of the remainder of the rules in this example for Problem 1.2. Note that **R** is also a linear space, where we make the identification $\mathbf{R} = \mathbf{R}_1$.

■

EXAMPLE 1.2 Consider *unitary space* $\mathbf{C}_n$. Vectors in the space are given by (1.1) and (1.2), where α_k and β_k, $k = 1, 2, \ldots, n$ are in $\mathbf{C}$. Addition and multiplication are defined by (1.3) and (1.4) where $\alpha \in \mathbf{C}$. Proof that $\mathbf{C}_n$ is a linear space follows the same lines as in Example 1.1. Note that $\mathbf{C}$ is a linear space, where we make the identification $\mathbf{C} = \mathbf{C}_1$.

■

EXAMPLE 1.3 Consider $\mathcal{C}(0,1)$, the space of real-valued functions continuous on the interval (0, 1). For f and g in $\mathcal{C}(0,1)$ and $\alpha \in \mathbf{R}$, we define addition and multiplication as follows:

$$(f + g)(\xi) = f(\xi) + g(\xi) \tag{1.6}$$

$$(\alpha f)(\xi) = \alpha f(\xi) \tag{1.7}$$

for all $\xi \in (0, 1)$. If we assume prior establishment of the rules for addition of two real-valued functions and multiplication of a real-valued function by a real scalar, it is easy to establish that $\mathcal{C}(0,1)$ is a linear space by showing that the rules in I and II are satisfied. For example, for addition rule d,

$$\begin{aligned}(f + g)(\xi) &= f(\xi) + g(\xi) \\ &= g(\xi) + f(\xi) \\ &= (g + f)(\xi)\end{aligned} \tag{1.8}$$

We leave the completion of the proof for Problem 1.3.

■

In ordinary vector analysis over two or three spatial coordinates, we are often concerned with vectors that are parallel (collinear). This concept can be generalized in an abstract linear space. Let $x_1, x_2, \ldots, x_n$ be elements of a set of vectors in $\mathcal{S}$. The vectors are *linearly dependent* if there exist $\alpha_k \in \mathbf{F}$, $k = 1, 2, \ldots, n$, not all zero, such that

$$\sum_{k=1}^{n} \alpha_k x_k = \mathbf{0} \tag{1.9}$$

If the only way to satisfy (1.9) is $\alpha_k = 0$, $k = 1, 2, \ldots, n$, then the elements x_k are *linearly independent*. The sum

$$\sum_{k=1}^{n} \alpha_k x_k$$

is called a *linear combination* of the vectors x_k.

EXAMPLE 1.4 In $\mathbf{R}_2$, let $x_1 = (1, 3),\ x_2 = (2, 6)$. We test x_1 and x_2 for linear dependence. We form

$$\mathbf{0} = (0, 0) = \alpha_1(1, 3) + \alpha_2(2, 6) = (\alpha_1 + 2\alpha_2, 3\alpha_1 + 6\alpha_2)$$

from which we conclude that

$$\alpha_1 + 2\alpha_2 = 0$$
$$3\alpha_1 + 6\alpha_2 = 0$$

These two equations are consistent and yield $\alpha_1 = -2\alpha_2$. Certainly, $\alpha_1 = \alpha_2 = 0$ satisfies this equation, but there is also an infinite number of nonzero possibilities. The vectors are therefore linearly dependent. Indeed, the reader can easily make a sketch to show that x_1 and x_2 are collinear.

■

EXAMPLE 1.5 In $\mathcal{C}(0,1)$, let a set of vectors be defined by $f_k(\xi) = \sqrt{2}\sin k\pi\xi$, $k = 1, 2, \ldots, n$. We test the vectors f_k for linear dependence. We form

$$\sum_{k=1}^{n} \alpha_k \sqrt{2}\sin k\pi\xi = 0 \tag{1.10}$$

where $\alpha_k \in \mathbf{R}$. The f_k, defined above, form an *orthonormal* set on $\xi \in (0, 1)$. That is,

$$\int_0^1 f_m f_k d\xi = \begin{cases} 0, & k \neq m \\ 1, & k = m \end{cases} \tag{1.11}$$

Multiplication of both sides of (1.10) by $\sqrt{2}\sin m\pi\xi,\ m = 1, 2, \ldots, n$ and integration over (0,1) give, with the help of (1.11), $\alpha_m = 0,\ m = 1, 2, \ldots, n$. The vectors f_k are therefore linearly independent.

■

In Example 1.5, we note that the elements f_k are finite in number. We recognize them as a finite subset of the countably infinite number of elements in the Fourier sine series $f_k,\ k = 1, 2, \ldots$. A question arises concerning the linear independence of sets containing a countably infinite number of vectors. Let $x_1, x_2 \ldots$ be an infinite set of vectors in $\mathcal{S}$. The vectors are linearly independent if every finite subset of the vectors is linearly independent. In Example 1.5, this requirement is realized, so that the infinite set of elements present in the Fourier sine series is linearly independent.

In an abstract linear space $\mathcal{S}$, it would be helpful to have a measure of how many and what sort of vectors describe the space. A linear space $\mathcal{S}$ has *dimension n* if it possesses a set of n independent vectors and if every set of $n+1$ vectors is dependent. If for every positive integer k we can find k independent vectors in $\mathcal{S}$, then $\mathcal{S}$ has infinite dimension. The set $x_1, x_2, \ldots, x_n$ is a *basis* for $\mathcal{S}$ provided that the vectors in the set are linearly independent, and provided that every $x \in \mathcal{S}$ can be written as a linear combination of the x_k, viz.

$$x = \sum_{k=1}^{n} \alpha_k x_k \tag{1.12}$$

The representation with respect to a given basis is unique. If it were not, then, in addition to the representation in (1.12), there would exist $\beta_k \in \mathbf{F}$, $k = 1, 2, \ldots, n$ such that

$$x = \sum_{k=1}^{n} \beta_k x_k \tag{1.13}$$

Subtraction of (1.13) from (1.12) yields

$$\mathbf{0} = \sum_{k=1}^{n} (\alpha_k - \beta_k) x_k \tag{1.14}$$

Since the x_k are linearly independent, we must have

$$\alpha_k - \beta_k = 0, \qquad k = 1, 2, \ldots, n \tag{1.15}$$

which proves uniqueness with respect to a given basis. Finally, if $\mathcal{S}$ is n-dimensional, any set of n linearly independent vectors $x_1, x_2, \ldots, x_n$ forms a basis. Indeed, let $x \in \mathcal{S}$. By the definition of dimension, the set $x, x_1, x_2, \ldots, x_n$ is linearly dependent, and therefore,

$$\alpha x + \sum_{k=1}^{n} \alpha_k x_k = \mathbf{0} \tag{1.16}$$

where we must have $\alpha \neq 0$. Dividing by α gives

$$x = \sum_{k=1}^{n} \left(\frac{-\alpha_k}{\alpha} \right) x_k \tag{1.17}$$

Therefore, the set $x_1, x_2, \ldots, x_n$ is a basis.

EXAMPLE 1.6 Consider Euclidean space $\mathbf{R}_n$. We shall show that the vectors $e_1 = (1, 0, \ldots, 0), e_2 = (0, 1, \ldots, 0), \ldots, e_n = (0, 0, \ldots, 1)$ satisfy the two requirements for a basis. First, the set $e_1, \ldots, e_n$ is independent (Problem 1.8). Second, if $a \in \mathbf{R}_n$,

$$\begin{aligned} a &= (\alpha_1, \alpha_2, \ldots, \alpha_n) \\ &= \alpha_1(1, 0, \ldots, 0) + \alpha_2(0, 1, \ldots, 0) + \cdots + \alpha_n(0, 0, \ldots, 1) \\ &= \alpha_1 e_1 + \alpha_2 e_2 + \cdots + \alpha_n e_n \end{aligned} \tag{1.18}$$

Therefore, any vector in the space can be expressed as a linear combination of the e_k. A special case of this result is obtained by considering Euclidean space $\mathbf{R}_3$. The vectors $e_1 = (1, 0, 0), e_2 = (0, 1, 0), e_3 = (0, 0, 1)$ are a basis. These vectors are perhaps best known as the unit vectors associated with the Cartesian coordinate system.

■

EXAMPLE 1.7 It would be consistent with notation if the dimension of $\mathbf{R}_n$ were, in fact, n. We now show that both requirements for dimension n are satisfied. First, since we have established an n-term basis for $\mathbf{R}_n$ in Example 1.6, the space has a set of n independent vectors. Second, we must show that *any* set of $n + 1$ vectors is dependent. Let $a_1, a_2, \ldots, a_n, a_{n+1}$ be an arbitrary set of $n + 1$ vectors in $\mathbf{R}_n$. We form the expression

$$\sum_{m=1}^{n+1} \gamma_m a_m = \mathbf{0} \tag{1.19}$$

where we must show that there exist $\gamma_m \in \mathbf{R}$, $m = 1, 2, \ldots, n + 1$, not all zero, such that (1.19) is satisfied. We express each of the members of the arbitrary set as a linear combination of the basis vectors, viz.

$$a_m = \sum_{k=1}^{n} \alpha_k^{(m)} e_k, \qquad m = 1, 2, \ldots, n + 1 \tag{1.20}$$

Substitution of (1.20) into (1.19) and interchanging the order of the summations gives

$$\sum_{k=1}^{n} \left(\sum_{m=1}^{n+1} \gamma_m \alpha_k^{(m)} \right) e_k = \mathbf{0} \tag{1.21}$$

Since the e_k are linearly independent,

$$\sum_{m=1}^{n+1} \gamma_m \alpha_k^{(m)} = \mathbf{0}, \qquad k = 1, 2, \ldots, n \tag{1.22}$$

Expression (1.22) is a homogeneous set of n linear equations in $n+1$ unknowns. The set is *underdetermined*, and as a result, always has a nontrivial solution [1]. There is therefore at least one nonzero coefficient among γ_m, $m = 1, 2, \ldots, n+1$. The result is that the arbitrary set $a_1, \ldots, a_n, a_{n+1}$ is linearly dependent and the dimension of $\mathbf{R}_n$ is n.

■

1.3 INNER PRODUCT SPACE

A linear space $\mathcal{S}$ is a *complex inner product space* if for every ordered pair (x, y) of vectors in $\mathcal{S}$, there exists a unique scalar in $\mathbf{C}$, symbolized $\langle x, y\rangle$, such that:

a. $\langle x, y\rangle = \overline{\langle y, x\rangle}$
b. $\langle x + y, z\rangle = \langle x, z\rangle + \langle y, z\rangle$
c. $\langle \alpha x, y\rangle = \alpha\langle x, y\rangle,\ \alpha \in \mathbf{C}$
d. $\langle x, x\rangle \geq 0$, with equality if and only if $x = \mathbf{0}$

In a, the overbar indicates complex conjugate. Similar to the above is the *real inner product space*, which we produce by eliminating the overbar in a and requiring in c that α be in $\mathbf{R}$. For the remainder of this section, we shall assume the complex case. We leave the reader to make the necessary specialization to the real inner product.

EXAMPLE 1.8 We show from the definition of complex inner product space that

$$\langle \mathbf{0}, y\rangle = 0 \tag{1.23}$$

Indeed, the result follows immediately if we substitute $\alpha = 0$ in rule c above.

■

EXAMPLE 1.9 Given the rules for the complex inner product in a–d, the following result holds:

$$\langle x, \alpha y\rangle = \bar{\alpha}\langle x, y\rangle \tag{1.24}$$

Indeed,

$$\begin{aligned} \langle x, \alpha y\rangle &= \overline{\langle \alpha y, x\rangle} \\ &= \overline{\alpha\langle y, x\rangle} \\ &= \bar{\alpha}\ \overline{\langle y, x\rangle} \\ &= \bar{\alpha}\langle x, y\rangle \end{aligned}$$

■

EXAMPLE 1.10 Given the rules for the inner product space, we may show that

$$\langle \sum_{k=1}^{n} \alpha_k x_k, y \rangle = \sum_{k=1}^{n} \alpha_k \langle x_k, y \rangle \tag{1.25}$$

The proof is left for Problem 1.9.

■

EXAMPLE 1.11 In the space $\mathbf{C}_n$, with a and b defined in Example 1.2, define an inner product by

$$\langle a, b \rangle = \sum_{k=1}^{n} \alpha_k \bar{\beta}_k \tag{1.26}$$

Then, $\mathbf{C}_n$ is a complex inner product space. To prove this, we must show that rules a–d for the complex inner product space are satisfied. For rule d, there are three parts to prove. First, we show that the inner product $\langle a, a \rangle$ is nonnegative. Indeed,

$$\langle a, a \rangle = \sum_{k=1}^{n} |\alpha_k|^2 \geq 0$$

Second, we show that $\langle a, a \rangle = 0$ implies that $a = \mathbf{0}$. We have

$$0 = \langle a, a \rangle = \sum_{k=1}^{n} |\alpha_k|^2$$

Since all the terms in the sum are nonnegative, $\alpha_k = 0$, $k = 1, 2, \ldots, n$, and therefore $a = \mathbf{0}$. Third, we must show that $a = \mathbf{0}$ implies $\langle a, a \rangle = 0$. We leave this for the reader. We also leave the reader to demonstrate that rules a–c for the inner product space are satisfied.

■

EXAMPLE 1.12 Let f and g be two vectors in $C(\alpha, \beta)$. Define an inner product by

$$\langle f, g \rangle = \int_{\alpha}^{\beta} f(\xi) g(\xi) d\xi \tag{1.27}$$

Then, $C(\alpha, \beta)$ is a real inner product space. We leave the proof for Problem 1.10.

■

One of the most important inequalities in linear analysis follows from the basic rules for the complex inner product space. The *Cauchy–Schwarz–Bunjakowsky inequality* is given by

$$|\langle x, y \rangle| \leq \sqrt{\langle x, x \rangle} \sqrt{\langle y, y \rangle} \tag{1.28}$$

Herein, we refer to (1.28) as the *CSB* inequality. To prove the CSB inequality, we first note that for $|\langle x, y\rangle| = 0$, there is nothing to prove. We may therefore assume $y \neq 0$, with the result $\langle y, y\rangle \neq 0$, and define

$$\alpha = \frac{\langle x, y\rangle}{\langle y, y\rangle}$$

from which we have the result

$$\begin{aligned}\frac{|\langle x, y\rangle|^2}{\langle y, y\rangle} &= \frac{\langle x, y\rangle\langle y, x\rangle}{\langle y, y\rangle}\\ &= \alpha\langle y, x\rangle\\ &= \bar{\alpha}\langle x, y\rangle\\ &= |\alpha|^2\langle y, y\rangle\end{aligned} \tag{1.29}$$

With the help of rule d, we form

$$\begin{aligned}0 \leq \langle x - \alpha y, x - \alpha y\rangle &= \langle x, x\rangle + |\alpha|^2\langle y, y\rangle - \bar{\alpha}\langle x, y\rangle - \alpha\langle y, x\rangle\\ &= \langle x, x\rangle - \frac{|\langle x, y\rangle|^2}{\langle y, y\rangle}\end{aligned}$$

from which the result in (1.28) follows.

Two concepts used throughout this book involve the notions of orthogonality and orthonormality. The concepts are generalizations of the ideas introduced in Example 1.5. Two vectors x and y are *orthogonal* if

$$\langle x, y\rangle = 0 \tag{1.30}$$

The set z_k, $k = 1, 2, \ldots$ is an orthogonal set if, for all members of the set,

$$\langle z_i, z_j\rangle = 0, \qquad i \neq j \tag{1.31}$$

The set is an *orthonormal* set if

$$\langle z_i, z_j\rangle = \delta_{ij} \tag{1.32}$$

where

$$\delta_{ij} = \begin{cases} 1, & i = j\\ 0, & i \neq j\end{cases} \tag{1.33}$$

An orthogonal set is called *proper* if it does not contain the zero vector. We can show that a proper orthogonal set of vectors is linearly independent. Indeed, we form

$$\sum_{k=1}^{n} \alpha_k z_k = \mathbf{0}$$

Taking the inner product of both sides with z_i gives

$$\langle \sum_{k=1}^{n} \alpha_k z_k, z_i \rangle = \langle \mathbf{0}, z_i \rangle$$

Using (1.25) and (1.31), we obtain

$$\alpha_i \langle z_i, z_i \rangle = 0$$

from which we conclude that $\alpha_i = 0, i = 1, 2, \ldots, n$ and the set is linearly independent. Further, if the index n is arbitrary, the countably infinite set $z_k, k = 1, 2, \ldots$, is linearly independent.

1.4 NORMED LINEAR SPACE

A linear space $\mathcal{S}$ is a *normed linear space* if, for every vector $x \in \mathcal{S}$, there is assigned a unique number $\|x\| \in \mathbf{R}$ such that the following rules apply:

a. $\|x\| \geq 0$, with equality if and only if $x = \mathbf{0}$
b. $\|\alpha x\| = |\alpha| \, \|x\|, \alpha \in \mathbf{F}$
c. $\|x_1 + x_2\| \leq \|x_1\| + \|x_2\|$ (triangle inequality)

Although there are many possible definitions of norms, we use exclusively the *norm induced by the inner product*, defined by

$$\|x\| = \sqrt{\langle x, x \rangle} \tag{1.34}$$

Using (1.34), we find that the CSB inequality in (1.28) can be written

$$|\langle x, y \rangle| \leq \|x\| \, \|y\| \tag{1.35}$$

It is easy to show that the norm defined by (1.34) meets the requirements in rules a, b, and c above. We leave the reader to show that a and b are satisfied. For c, for x and y in $\mathcal{S}$, we have

$$\begin{aligned}
\|x + y\|^2 &= \langle x + y, x + y \rangle \\
&= \langle x, x \rangle + \langle y, y \rangle + \langle x, y \rangle + \langle y, x \rangle \\
&= \|x\|^2 + \|y\|^2 + 2\text{Re}\langle x, y \rangle
\end{aligned}$$

Since the real part of a complex number is less than or equal to its magnitude,

$$\|x + y\|^2 \leq \|x\|^2 + \|y\|^2 + 2|\langle x, y \rangle|$$

Using the CSB inequality, we obtain

$$\|x + y\|^2 \leq \|x\|^2 + \|y\|^2 + 2\|x\| \, \|y\|$$

Taking the square root of both sides yields the result in c.

EXAMPLE 1.13 From the basic rules for the norm and the definition in (1.34), we can show that

$$\|x + y\|^2 + \|x - y\|^2 = 2\left(\|x\|^2 + \|y\|^2\right) \tag{1.36}$$

Indeed,

$$\begin{aligned}\|x + y\|^2 + \|x - y\|^2 &= \langle x + y, x + y\rangle + \langle x - y, x - y\rangle \\ &= 2\langle x, x\rangle + 2\langle y, y\rangle\end{aligned}$$

from which the result in (1.36) follows.

■

EXAMPLE 1.14 For unitary space $\mathbf{C}_n$, with inner product defined by (1.26), the norm of a vector a in the space is easily found to be

$$\|a\| = \sqrt{\sum_{k=1}^{n} |\alpha_k|^2} \tag{1.37}$$

Unitary space is therefore a normed linear space.

■

EXAMPLE 1.15 For the real linear space $\mathcal{C}(\alpha, \beta)$, with inner product defined by (1.27), the norm of a vector f in the space is

$$\|f\| = \sqrt{\int_{\alpha}^{\beta} f^2(\xi)d\xi} \tag{1.38}$$

The space $\mathcal{C}(\alpha, \beta)$ is therefore a normed linear space.

■

One of the useful consequences of the normed linear space is that it provides a measure of the "closeness" of one vector to another. We note from rule a that $\|x - y\| = 0$ if and only if $x = y$. Therefore, closeness can be indicated by the relation $\|x - y\| < \epsilon$. This observation brings us to the notion of convergence. Among the many forms of convergence, there are two forms whose relationship is crucial to placing firm "boundaries" on the linear space. The type of boundary we seek is one that assures that the limit of a sequence in the linear space also is contained in the space.

In a normed linear space $\mathcal{S}$, a sequence of vectors $\{x_k\}_{k=1}^{\infty}$ *converges* to a vector $x \in \mathcal{S}$ if, given an $\epsilon > 0$, there exists a number N such that $\|x - x_k\| < \epsilon$ whenever $k > N$. We write $x_k \to x$ or

$$\lim_{k\to\infty} x_k = x \tag{1.39}$$

Note that if $x_k \to x$, $\|x - x_k\| \to 0$.

Fundamental to studies of approximation of one vector by another vector, to be studied later in this chapter, is the notion of *continuity of the inner product*. We show that if $\{x_k\}_{k=1}^{\infty}$ is a sequence in $\mathcal{S}$ converging to $x \in \mathcal{S}$, then

$$\langle x_k, h\rangle \to \langle x, h\rangle \tag{1.40}$$

where h is any vector in $\mathcal{S}$. To prove (1.40), it is sufficient to show that

$$\langle x_k, h\rangle - \langle x, h\rangle \to 0$$

or

$$\langle x_k - x, h\rangle \to 0 \tag{1.41}$$

By the form of the CSB inequality in (1.35), we have

$$|\langle x_k - x, h\rangle|^2 \le \|x_k - x\|^2\|h\|^2$$

But, since $x_k \to x$,

$$\|x_k - x\| \to 0$$

so that (1.41) is verified. We remark that another useful way of writing (1.40) is as follows:

$$\lim_{k\to\infty}\langle x_k, h\rangle = \langle \lim_{k\to\infty} x_k, h\rangle \tag{1.42}$$

This relationship indicates that, given x_k, the order of application of the limit and the inner product with h can be interchanged.

In $\mathcal{S}$, a sequence $\{x_k\}_{k=1}^{\infty}$ *converges in the Cauchy sense* if, given an $\epsilon > 0$, there exists a number N such that $\|x_m - x_n\| < \epsilon$ whenever $\min(m, n) > N$. We write

$$\lim_{m,n\to\infty} \|x_m - x_n\| = 0 \tag{1.43}$$

We can show that convergence implies Cauchy convergence. Indeed, let $x \in \mathcal{S}$ be defined as the limit of a sequence, as in (1.39). Then, by the triangle inequality,

$$\|x_m - x_n\| \leq \|x_m - x\| + \|x - x_n\|$$

Since $x_k \to x$, there exists a number N such that for $n > N$

$$\|x - x_n\| \leq \frac{\epsilon}{2}$$

and therefore, for $\min(m, n) > N$,

$$\|x_m - x_n\| \leq \epsilon$$

which proves the assertion. Unfortunately, the converse is not always true. The interpretation is that it is possible for two members of the sequence to become arbitrarily close without the sequence itself approaching a limit in $\mathcal{S}$. A normed linear space is said to be *complete* if every Cauchy sequence in the space converges to a vector in the space. The concept of completeness is an important one in what is to follow. Although it is beyond the scope of this book to include a detailed treatment, we shall give a brief discussion.

In real analysis, the space of *rational numbers* is defined [2] as those numbers that can be written as p/q, where p and q are integers. It is a standard exercise [3],[4] to produce a sequence of rational numbers that has the Cauchy property and yet fails to converge in the space. (We consider an example in Problem 1.15.) This incompleteness is caused by the fact that in between two rational numbers, no matter how close, is an infinite number of irrational numbers; often, a sequence of rationals can converge to an irrational. The solution to this problem is a procedure due to Cantor [5] whereby the irrationals are appended to the rationals in such a manner so as to produce a complete linear space called the space of real numbers **R**. We shall assume henceforth that **R** is complete, and direct the reader to the literature in real analysis for details.

EXAMPLE 1.16 We can show that Euclidean space $\mathbf{R}_n$ is complete. For vectors a and b in the space, defined by (1.1) and (1.2), we define an inner product by

$$\langle a, b \rangle = \sum_{k=1}^{n} \alpha_k \beta_k \tag{1.44}$$

Let a_m, $m = 1, 2, \ldots$ be a Cauchy sequence in $\mathbf{R}_n$, where

$$a_m = \left(\alpha_1^{(m)}, \alpha_2^{(m)}, \ldots, \alpha_n^{(m)}\right)$$

Then,

$$\|a_m - a_p\| = \left\{ \sum_{k=1}^{n} \left[\alpha_k^{(m)} - \alpha_k^{(p)} \right]^2 \right\}^{\frac{1}{2}} \leq \epsilon$$

for $\min(m, p) > N$. Since all the terms in the sum are nonnegative, we must have

$$\left| \alpha_k^{(m)} - \alpha_k^{(p)} \right| \leq \epsilon, \qquad k = 1, 2, \ldots, n$$

for $\min(m, p) > N$. Since the space of real numbers **R** is complete, then as $m \to \infty$,

$$\alpha_k^{(m)} \to \alpha_k, \qquad k = 1, 2, \ldots, n$$

and therefore $a_m \to a$.

■

EXAMPLE 1.17 We can show that the normed linear space $\mathcal{C}(\alpha, \beta)$ with norm given by (1.38) is incomplete. We shall consider a well-known [6],[7] Cauchy sequence that fails to converge to a vector in the space. Without loss of generality, let (α, β) be (-1,1). Consider the sequence

$$f_k(\xi) = \begin{cases} 0, & -1 \leq \xi \leq 0 \\ k\xi, & 0 \leq \xi \leq \frac{1}{k} \\ 1, & \frac{1}{k} \leq \xi \leq 1 \end{cases} \tag{1.45}$$

where $k = 1, 2, \ldots$. This sequence is continuous, and therefore is in the linear space $\mathcal{C}$(-1,1). We show that this sequence is Cauchy. We form the difference between two members of the sequence. For $m > k$,

$$f_m - f_k = \begin{cases} 0, & -1 \leq \xi \leq 0 \\ (m-k)\xi, & 0 \leq \xi \leq \frac{1}{m} \\ (1 - k\xi), & \frac{1}{m} \leq \xi \leq \frac{1}{k} \\ 0, & \frac{1}{k} \leq \xi \leq 1 \end{cases}$$

We display the two sequence members f_k and f_m in Fig. 1-1. Note that the difference $f_m - f_k$ is always less than unity. It follows that unity is an upper bound on $(f_m - f_k)^2$. We therefore must have

$$\|f_m - f_k\|^2 = \int_{-1}^{1} (f_m(\xi) - f_k(\xi))^2 d\xi \leq \frac{1}{k}$$

Although the above result has been obtained for $m > k$, an interchange of m and k gives the general result

$$\|f_m - f_k\|^2 \le \max\left(\frac{1}{m}, \frac{1}{k}\right)$$

$$\lim_{m,k\to\infty} \|f_m - f_k\| = 0$$

which proves that the sequence is Cauchy. However, it is apparent that the sequence f_k converges to the Heaviside function $H(\xi)$, defined by

$$H(\xi) = \begin{cases} 0, & \xi < 0 \\ 1, & \xi > 0 \end{cases} \tag{1.46}$$

But, $H(\xi) \notin \mathcal{C}(-1,1)$. Therefore, the space $\mathcal{C}(\alpha, \beta)$ is not complete.

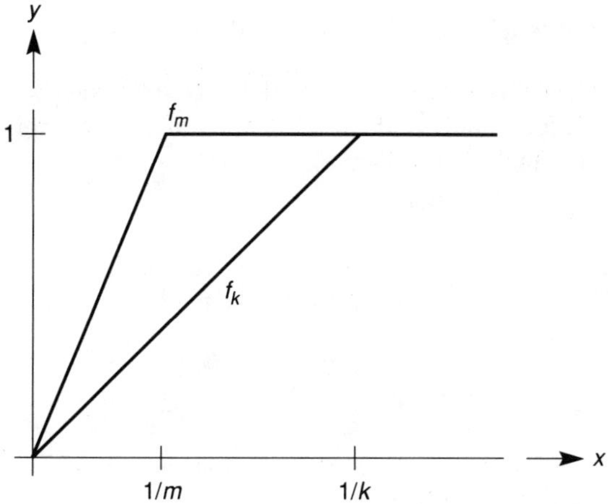

Fig. 1-1 Two members f_k and f_m of sequence in (1.45) for the case $m > k$.

■

1.5 HILBERT SPACE

A linear space is a *Hilbert space* if it is complete in the norm induced by the inner product. Therefore, in any Hilbert space, Cauchy convergence implies convergence. From Example 1.16, Euclidean space $\mathbf{R}_n$ is complete in the norm induced by the inner product in (1.44). Therefore, $\mathbf{R}_n$ is a Hilbert space. In a similar manner, it can be shown that unitary space $\mathbf{C}_n$ is complete. However, from Example 1.17, $\mathcal{C}(\alpha, \beta)$ is incomplete. In a manner similar to the completion of the space of rational numbers, $\mathcal{C}$ can be completed. The result is $\mathcal{L}_2(\alpha, \beta)$, the Hilbert space of real functions $f(\xi)$ square integrable on the interval (α, β), viz.

$$\int_\alpha^\beta f^2(\xi)d\xi < \infty \tag{1.47}$$

In (1.47), the integration is to be understood in the Lebesgue sense [8]. Although the Lebesgue theory is essential to the understanding of the proof of completeness, in this book such proofs will be omitted. For discussions of the issues involved, the reader is directed to [9],[10].

In linear analysis, we are often concerned with subsets of vectors in a linear space. One such subset is called a *linear manifold.* If $\mathcal{S}$ is a linear space and $\alpha, \beta \in \mathbf{F}$, then $\mathcal{M}$ is a linear manifold in $\mathcal{S}$, provided that $\alpha x + \beta y$ are in $\mathcal{M}$ whenever x and y are in $\mathcal{M}$. It is easy to show that $\mathcal{M}$ is a linear space. The proof is left for Problem 1.16.

EXAMPLE 1.18 In $\mathbf{R}_2$, the set of all vectors in the first quadrant is *not* a linear manifold. Indeed, for x and y in the first quadrant, it is easy to find $\alpha, \beta \in \mathbf{R}$ such that $\alpha x + \beta y$ is not in the first quadrant.

■

EXAMPLE 1.19 Let $x_k, k = 1, 2, \ldots, n$ be a linearly independent sequence of vectors in the Hilbert space $\mathcal{H}$. Define $\mathcal{M}$ to be the set of all linear combinations of the n vectors, viz.

$$\sum_{k=1}^{n} \alpha_k x_k$$

Then, $\mathcal{M}$ is a linear manifold. We leave the proof for Problem 1.17. $\mathcal{M}$ is called the linear manifold *generated* by $x_k, k = 1, 2, \ldots, n$.

■

The results in Example 1.19 raise an interesting issue. Can the same linear manifold be generated by more than one sequence of vectors? We shall show that indeed this is the case. We consider the *Gram–Schmidt orthogonalization process.* Let $\{x_1, \ldots, x_n\}$ be a linearly independent sequence of vectors generating the linear manifold $\mathcal{M} \subset \mathcal{H}$. The Gram–Schmidt process is a constructive procedure for generating an orthonormal sequence $\{e_1, \ldots, e_n\}$ from the independent sequence. To begin, let

$$z_1 = x_1 \tag{1.48}$$

and

$$e_1 = \frac{z_1}{\|z_1\|} \tag{1.49}$$

The next member of the sequence is generated by

$$z_2 = x_2 - \langle x_2, e_1 \rangle e_1 \tag{1.50}$$

and

$$e_2 = \frac{z_2}{\|z_2\|} \tag{1.51}$$

For the third member, we form

$$z_3 = x_3 - \langle x_3, e_1 \rangle e_1 - \langle x_3, e_2 \rangle e_2 \tag{1.52}$$

and

$$e_3 = \frac{z_3}{\|z_3\|} \tag{1.53}$$

This process continues until the final member of the sequence is produced by

$$z_n = x_n - \sum_{k=1}^{n-1} \langle x_n, e_k \rangle e_k \tag{1.54}$$

and

$$e_n = \frac{z_n}{\|z_n\|} \tag{1.55}$$

We leave the reader to show that the sequence $\{e_1, \ldots, e_n\}$ possesses the orthonormal property. In addition, each e_k, $k = 1, \ldots, n$ is a linear combination of $x_1, \ldots, x_k$. We conclude that any linear combination

$$\sum_{k=1}^{n} \alpha_k e_k$$

is also a linear combination

$$\sum_{k=1}^{n} \beta_k x_k$$

The original sequence and the orthornormal sequence obtained from it therefore generate the same linear manifold.

EXAMPLE 1.20 Given the sequence $\{1, \tau, \tau^2, \ldots\} \in \mathcal{L}_2(-1, 1)$, we use the Gram–Schmidt procedure to produce an orthonormal sequence. Indeed, we define an inner product by

$$\langle f, g \rangle = \int_{-1}^{1} f(\tau) g(\tau) d\tau$$

Then,

$$z_1(\tau) = 1$$
$$e_1(\tau) = 1/\|1\| = 1/\sqrt{2}$$
$$z_2(\tau) = \tau - \langle \tau, 1/\sqrt{2}\rangle(1/\sqrt{2}) = \tau$$
$$e_2(\tau) = \sqrt{3/2}\,\tau$$
$$z_3(\tau) = \tau^2 - 1/3$$
$$e_3(\tau) = \sqrt{45/8}\,(\tau^2 - 1/3)$$

This process continues for as many terms in the orthonormal sequence as we wish to calculate. We remark that the members of the sequence so produced are proportional to the orthogonal sequence of *Legendre polynomials*, whose first few members are

$$P_0(\tau) = 1$$
$$P_1(\tau) = \tau$$
$$P_2(\tau) = \frac{1}{2}(3\tau^2 - 1)$$
$$P_3(\tau) = \frac{1}{2}(5\tau^3 - 3\tau)$$
$$P_4(\tau) = \frac{1}{8}(35\tau^4 - 30\tau^2 + 3)$$
$$P_5(\tau) = \frac{1}{8}(63\tau^5 - 70\tau^3 + 15\tau)$$

The Legendre functions, orthogonal but not orthonormal, are constructed in such a way that $P_n(\pm 1) = \pm 1$.

■

We next discuss a characteristic associated with linear manifolds that plays a central role in approximation theory. A linear manifold $\mathcal{M}$ is said to be *closed* if it contains the limits of all sequences that can be constructed from the members of $\mathcal{M}$. It is easy to demonstrate that not all linear manifolds are closed. For example, the space $\mathcal{C}(\alpha, \beta)$ is a linear manifold since a linear combination of two continuous functions is a continuous function. However, in Example 1.17, we have given a sequence of vectors in $\mathcal{C}$ that fails to converge to a vector in $\mathcal{C}$. The linear manifold $\mathcal{C}$ is therefore not closed.

An interesting result occurs if a closed linear manifold is contained in a Hilbert space. Specifically, if $\mathcal{H}$ is a Hilbert space and $\mathcal{M}$ is a closed linear manifold in $\mathcal{H}$, then $\mathcal{M}$ is a Hilbert space. Indeed, let $\{x_k\}_{k=1}^{\infty}$ be a Cauchy sequence in $\mathcal{M}$. Then, since $\mathcal{M}$ is contained in the Hilbert space

$\mathcal{H}$, $x_k \to x \in \mathcal{H}$. But $\mathcal{M}$ is closed, and therefore $x \in \mathcal{M}$. We conclude that $\mathcal{M}$ is a Hilbert space.

1.6 BEST APPROXIMATION

Within the structure of the Hilbert space, it is possible to generalize the concepts of approximation of vectors and functions. Let x be a vector in a Hilbert space $\mathcal{H}$ and let $\{z_k\}_{k=1}^m$ be an orthonormal set in $\mathcal{H}$. We form the sum

$$x_m = \sum_{k=1}^{m} \alpha_k z_k \tag{1.56}$$

This sum generates a linear manifold $\mathcal{M} \subset \mathcal{H}$. Different members of the linear manifold are produced by assigning various values to the sequence of coefficients $\{\alpha_k\}_{k=1}^m$. We should like to determine what choice of coefficients results in x_m being the "best" approximation to x. Specifically, let us make x_m "close" to x by adjusting the coefficients to minimize $\|x - x_m\|$. We expand the square of the norm as follows:

$$\begin{aligned}
\|x - x_m\|^2 &= \langle x - x_m, x - x_m \rangle \\
&= \langle x, x \rangle + \langle x_m, x_m \rangle - \langle x_m, x \rangle - \langle x, x_m \rangle \\
&= \|x\|^2 + \sum_{k=1}^{m} |\alpha_k|^2 - \sum_{k=1}^{m} \alpha_k \overline{\langle x, z_k \rangle} - \sum_{k=1}^{m} \overline{\alpha_k} \langle x, z_k \rangle
\end{aligned}$$

Completing the square, we obtain

$$\|x - x_m\|^2 = \|x\|^2 + \sum_{k=1}^{m} (\alpha_k - \langle x, z_k \rangle) \overline{(\alpha_k - \langle x, z_k \rangle)} - \sum_{k=1}^{m} |\langle x, z_k \rangle|^2 \tag{1.57}$$

Since the sum in (1.57) containing the coefficients α_k is nonnegative, the norm-squared (and hence the norm) is minimized by the choice

$$\alpha_k = \langle x, z_k \rangle, \qquad k = 1, 2, \ldots, m \tag{1.58}$$

Expression (1.58) defines the *Fourier coefficients* associated with orthonormal expansions. Note that once we have made the selection given in (1.58), we can define an error vector e_m by

$$\begin{aligned}
e_m &= x - x_m \\
&= x - \sum_{k=1}^{m} \langle x, z_k \rangle z_k
\end{aligned} \tag{1.59}$$

Taking the inner product with any member z_j of the orthonormal sequence, we find that

$$\langle e_m, z_j \rangle = \langle x, z_j \rangle - \sum_{k=1}^{m} \langle x, z_k \rangle \langle z_k, z_j \rangle = 0, \qquad j = 1, 2, \ldots, m \quad (1.60)$$

Since x_m is a linear combination of members of the sequence $\{z_j\}_{j=1}^{m}$, we must have

$$\langle e_m, x_m \rangle = 0 \quad (1.61)$$

We summarize these results as follows:

a. The vector $x \in \mathcal{H}$ has been decomposed into a vector $x_m \in \mathcal{M} \subset \mathcal{H}$ plus an error vector e_m, viz.

$$x = x_m + e_m \quad (1.62)$$

b. The error vector e_m is orthogonal to the approximation vector x_m.

EXAMPLE 1.21 We consider the Fourier sine series in Example 1.5. Let $f(\xi) \in \mathcal{L}_2(0, 1)$ with inner product

$$\langle f, g \rangle = \int_0^1 f(\xi) g(\xi) d\xi \quad (1.63)$$

Let $\{\sqrt{2} \sin k\pi\xi\}_{k=1}^{m}$ be an orthonormal set generating a linear manifold $\mathcal{M} \subset \mathcal{H}$. Then, by (1.56) and (1.58), the "best" approximating function $f_m \in \mathcal{M}$ to $f(\xi)$ is given by

$$f_m(\xi) = \sum_{k=1}^{m} \alpha_k \sqrt{2} \sin k\pi\xi \quad (1.64)$$

where

$$\alpha_k = \int_0^1 f(\eta) \sqrt{2} \sin k\pi\eta d\eta \quad (1.65)$$

which is the classical Fourier result.

■

The above results for the approximation of a vector $x \in \mathcal{H}$ by a vector $x_m \in \mathcal{M} \subset \mathcal{H}$ can be generalized. We shall need the concept of a manifold that is orthogonal to a given manifold. If $\mathcal{M}$ is a linear manifold, then the vector $e \in \mathcal{H}$ is a member of a set $\mathcal{M}^{\perp}$ if it is orthogonal to every vector in $\mathcal{M}$. The set $\mathcal{M}^{\perp}$ is a linear manifold since linear combinations of vectors in $\mathcal{M}^{\perp}$ are also orthogonal to vectors in $\mathcal{M}$. In fact, $\mathcal{M}^{\perp}$ is also closed.

Indeed, let $\{e_k\}_{k=1}^{\infty}$ be a sequence in $\mathcal{M}^{\perp}$ converging to a vector e in $\mathcal{H}$. We have

$$\langle e_k, x\rangle = 0$$

for all $x \in \mathcal{M}$. But, by continuity of the inner product,

$$\lim_{k\to\infty} \langle e_k, x\rangle = \langle e, x\rangle = 0$$

so that $e \in \mathcal{M}^{\perp}$, which is therefore closed. The closed linear manifold $\mathcal{M}^{\perp}$ is called the *orthogonal complement* to $\mathcal{M}$.

We now can produce a result called the *Projection Theorem*. Let x be any vector in the Hilbert space $\mathcal{H}$, and let $\mathcal{M} \subset \mathcal{H}$ be a *closed* linear manifold. Then, there is a unique vector $y_0 \in \mathcal{M} \subset \mathcal{H}$ closest to x in the sense that $\|x - y_0\| \leq \|x - y\|$ for all y in $\mathcal{M}$. Furthermore, the necessary and sufficient condition that y_0 is the unique minimizing vector is that $e = x - y_0$ is in $\mathcal{M}^{\perp}$. The proof of this important theorem is deferred to Appendix A.1 at the end of the chapter. The vector y_0 is called the *projection* of x onto $\mathcal{M}$. The vector e is called the projection of x onto $\mathcal{M}^{\perp}$. The ideas inherent to the projection theorem have a well-known interpretation in two- and three-dimensional vector spaces, as in the following example.

EXAMPLE 1.22 Let $a = (\alpha_1, \alpha_2)$ be any vector in $\mathbf{R}_2$. Let $\mathcal{M}$ be the set of all vectors b in $\mathbf{R}_2$ with second component equal to zero, viz.

$$b = (\beta_1, 0)$$

Since linear combinations of vectors in $\mathcal{M}$ are also in $\mathcal{M}$, the set $\mathcal{M}$ is a linear manifold. In addition, the manifold is closed. Indeed, let $b_1, b_2, \ldots$ be a sequence in $\mathcal{M}$ converging to $b \in \mathcal{H}$, where

$$b_k = \left(\beta_1^{(k)}, 0\right)$$

Since $b_k \to b \in \mathcal{H}$, $\beta_1^{(k)} \to \beta_1$. Therefore, $b \in \mathcal{M}$. By the projection theorem, among all vectors $b \in \mathcal{M}$, the vector $\hat{b}$ closest to a can be obtained from

$$\langle a - \hat{b}, b\rangle = 0$$

where

$$\hat{b} = (\hat{\beta}_1, 0)$$

Solving this equation, employing the usual definition of inner product for $\mathbf{R}_2$, we obtain

$$(\alpha_1 - \hat{\beta}_1)\beta_1 = 0$$

Since β_1 is arbitrary, $\hat{\beta}_1 = \alpha_1$. This result can be visualized (Fig. 1-2) by drawing the vector a and noting that the vector b lies along the x-axis. The "best" b is then obtained by dropping a perpendicular from the tip of a to the x-axis. The result is a vector $\hat{b}$, called the projection of a onto the x-axis. Note that the error vector e is such that it is orthogonal to any vector along the x-axis and that $a = \hat{b} + e$, as required by the projection theorem.

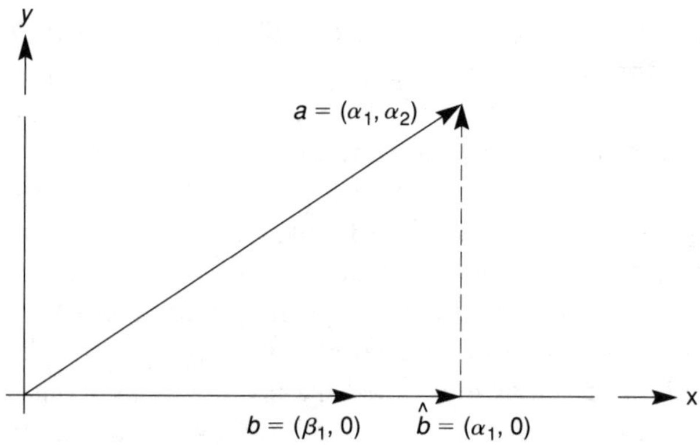

Fig. 1-2 Illustration of the projection theorem in $\mathbf{R}_2$, as given in Example 1.22.

■

In (1.56)–(1.58), we introduced best approximation in terms of orthonormal expansion functions generating a linear manifold. With the aid of the projection theorem, we next generalize the concept of best approximation to include expansion functions that are linearly independent but not necessarily orthogonal. Let $y \in \mathcal{H}$, and let $\{y_j\}_{j=1}^{M}$ be a linearly independent sequence of vectors in $\mathcal{H}$. We form the sum

$$\hat{y} = \sum_{j=1}^{M} \alpha_j y_j \tag{1.66}$$

We wish to approximate y with $\hat{y}$ by suitable choice of the coefficients α_j. We have already indicated in Example 1.19 (Problem 1.17) that linear combinations of the type in (1.66) form a linear manifold $\mathcal{M}$. In fact, since the limit of sequences of vectors in $\mathcal{M}$ must necessarily be in $\mathcal{M}$, the manifold is closed and therefore meets the requirements of the projection theorem. Since $y_j \in \mathcal{M}$, $j = 1, 2, \ldots, M$, the projection theorem gives

$$\langle y - \hat{y}, y_k \rangle = 0, \qquad k = 1, 2, \ldots, M \tag{1.67}$$

Substituting (1.66) and rearranging, we have

$$\sum_{j=1}^{M} \alpha_j \langle y_j, y_k \rangle = \langle y, y_k \rangle, \qquad k = 1, 2, \ldots, M \tag{1.68}$$

We write (1.68) in matrix form as follows:

$$\begin{bmatrix} \langle y_1, y_1 \rangle & \langle y_2, y_1 \rangle & \cdots & \langle y_M, y_1 \rangle \\ \langle y_1, y_2 \rangle & \langle y_2, y_2 \rangle & \cdots & \langle y_M, y_2 \rangle \\ \vdots & \vdots & \ddots & \vdots \\ \langle y_1, y_M \rangle & \langle y_2, y_M \rangle & \cdots & \langle y_M, y_M \rangle \end{bmatrix} \begin{bmatrix} \alpha_1 \\ \alpha_2 \\ \vdots \\ \alpha_M \end{bmatrix} = \begin{bmatrix} \langle y, y_1 \rangle \\ \langle y, y_2 \rangle \\ \vdots \\ \langle y, y_M \rangle \end{bmatrix} \tag{1.69}$$

Inversion of this matrix yields the coefficients α_j, $j = 1, 2, \ldots, M$. These coefficients then determine $\hat{y}$ in (1.66). The square matrix on the left-hand side of (1.69) is the *transpose* of a matrix called the *Gram matrix*. In addition to its appearance in best approximation, it also finds use in proofs of linear independence (Problem 1.21). Note that the result in (1.69) is a generalization of the Fourier coefficient result in (1.58). Indeed, for cases where the independent sequence y_j, $j = 1, 2, \ldots, M$ is orthogonal, the matrix in (1.69) diagonalizes. Inversion then produces the Fourier coefficient result.

One of the classic problems of algebra is the approximation of a function by a polynomial. This problem is easily cast as best approximation in the following example.

EXAMPLE 1.23 In the Hilbert space $\mathcal{L}_2(0, 1)$, consider the approximation of a function $f(\tau)$ by a polynomial. Let $\{\tau^{n-1}\}_{n=1}^{N}$ be a sequence in $\mathcal{L}_2(0, 1)$. The sequence is linearly independent. Indeed, by the fundamental theorem of algebra, the equation

$$\sum_{n=1}^{N} \beta_n \tau^{n-1} = 0$$

can have at most $N - 1$ roots. Therefore, the only solution valid for all $\tau \in (0, 1)$ is $\beta_n = 0, n = 1, 2, \ldots, N$. We wish to approximate $f(\tau)$ by

$$\hat{f}(\tau) = \sum_{n=1}^{N} \alpha_n \tau^{n-1} \tag{1.70}$$

All possible such linear combinations form a closed linear manifold so that the projection theorem applies. Comparing (1.66), we identify

$$y_n = \tau^{n-1} \tag{1.71}$$

Define an inner product for $\mathcal{L}_2(0, 1)$ by

$$\int_0^1 f(\tau)g(\tau)d\tau \tag{1.72}$$

We then have

$$\langle y_m, y_n \rangle = \int_0^1 \tau^{m-1}\tau^{n-1}d\tau = \frac{1}{m+n-1} \tag{1.73}$$

$$\langle y, y_m \rangle = \int_0^1 \tau^{m-1} f(\tau)d\tau \tag{1.74}$$

Substitution of (1.73) and (1.74) into (1.69) gives

$$\begin{bmatrix} 1 & \frac{1}{2} & \cdots & \frac{1}{N} \\ \frac{1}{2} & \frac{1}{3} & \cdots & \frac{1}{N+1} \\ \vdots & \vdots & \ddots & \vdots \\ \frac{1}{N} & \frac{1}{N+1} & \cdots & \frac{1}{2N-1} \end{bmatrix} \begin{bmatrix} \alpha_1 \\ \alpha_2 \\ \vdots \\ \alpha_N \end{bmatrix} = \begin{bmatrix} \int_0^1 f(\tau)d\tau \\ \int_0^1 \tau f(\tau)d\tau \\ \vdots \\ \int_0^1 \tau^{N-1} f(\tau)d\tau \end{bmatrix} \tag{1.75}$$

Inversion of this matrix equation yields the best approximation.

■

1.7 OPERATORS IN HILBERT SPACE

Consider the following transformation in $\mathbf{R}_3$:

$$\zeta_1 = \alpha_{11}\xi_1 + \alpha_{12}\xi_2 + \alpha_{13}\xi_3$$
$$\zeta_2 = \alpha_{21}\xi_1 + \alpha_{22}\xi_2 + \alpha_{23}\xi_3$$
$$\zeta_3 = \alpha_{31}\xi_1 + \alpha_{32}\xi_2 + \alpha_{33}\xi_3$$

Using the usual matrix notation, we let

$$z = [\,\zeta_1 \quad \zeta_2 \quad \zeta_3\,]^T$$

$$x = [\,\xi_1 \quad \xi_2 \quad \xi_3\,]^T$$

where T indicates matrix transpose. We then have

$$Ax = z$$

where

$$A = \begin{bmatrix} \alpha_{11} & \alpha_{12} & \alpha_{13} \\ \alpha_{21} & \alpha_{22} & \alpha_{23} \\ \alpha_{31} & \alpha_{32} & \alpha_{33} \end{bmatrix}$$

The solution is given formally by

$$x = A^{-1}z$$

The matrix operation is linear. Indeed, given x_1, x_2, z_1, and z_2, we have by ordinary matrix methods

$$A(\alpha x_1 + \beta x_2) = \alpha A x_1 + \beta A x_2 = \alpha z_1 + \beta z_2$$

The concepts of linearity and inversion for matrices can be generalized to linear operators in a Hilbert space.

An *operator* L is a mapping that assigns to a vector $x \in \mathcal{S}$ another vector $Lx \in \mathcal{S}$. We write

$$Lx = y \tag{1.76}$$

The *domain* of the operator L is the set of vectors x for which the mapping is defined. The *range* of the operator L is the set of vectors y resulting from the mapping. The operator L is *linear* if the mapping is such that for any x_1 and x_2 in the domain of L, the vector $\alpha_1 x_1 + \alpha_2 x_2$ is also in the domain and

$$L(\alpha_1 x_1 + \alpha_2 x_2) = \alpha_1 L x_1 + \alpha_2 L x_2 \tag{1.77}$$

A linear operator L with domain $\mathcal{D}_L \subset \mathcal{H}$ is *bounded* if there exists a real number γ such that

$$\|Lu\| \leq \gamma \|u\| \tag{1.78}$$

for all $u \in \mathcal{D}_L$.

EXAMPLE 1.24 Let $\mathbf{R}_\infty$ be the space of all vectors consisting of a countably infinite set of real numbers (components), viz.

$$a = (\alpha_1, \alpha_2, \alpha_3 \ldots) \tag{1.79}$$

where $\alpha_k \in \mathbf{R}$. If $b = (\beta_1, \beta_2, \beta_3, \ldots)$, define an inner product for the space by

$$\langle a, b \rangle = \sum_{k=1}^{\infty} \alpha_k \beta_k \tag{1.80}$$

Let the norm for the space be induced by the inner product. We restrict $\mathbf{R}_\infty$ to those vectors with finite norm. Define the *right shift* operator A_R in $\mathbf{R}_\infty$ by

$$A_R a = (0, \alpha_1, \alpha_2, \ldots)$$

The right shift operator A_R is linear. The proof is easy and is omitted. In addition, A_R is bounded. Indeed,

$$\|A_R a\| = \sqrt{\sum_{k=1}^{\infty} \alpha_k^2} = \|a\|$$

Therefore, the operator A_R is bounded in $\mathbf{R}_\infty$. Indeed, the least upper bound on γ is unity.

■

EXAMPLE 1.25 On the complex Hilbert space $\mathcal{L}_2(0, 1)$, we consider the following integral equation:

$$\int_0^1 u(\xi')k(\xi, \xi')d\xi' = f(\xi) \tag{1.81}$$

This equation can be written in operator notation as follows:

$$Lu = f$$

where L is the linear operator given by

$$L = \int_0^1 (\,\cdot\,)k(\xi, \xi')d\xi' \tag{1.82}$$

We shall show that the operator L is bounded if

$$\int_0^1 \int_0^1 |k(\xi, \xi')|^2 d\xi d\xi' < \infty \tag{1.83}$$

This property of the *kernel* $k(\xi, \xi')$ is called the *Hilbert–Schmidt* property, and the operator L it generates is called a *Hilbert–Schmidt* operator. To show that L is bounded, we form

$$\|Lu\|^2 = \int_0^1 |f(\xi)|^2 d\xi$$

where

$$\begin{aligned} |f(\xi)|^2 &= \left| \int_0^1 u(\xi')k(\xi, \xi')d\xi' \right|^2 \\ &\le \int_0^1 |u(\xi')|^2 d\xi' \int_0^1 |k(\xi, \xi')|^2 d\xi' \\ &= \|u\|^2 \int_0^1 |k(\xi, \xi')|^2 d\xi' \end{aligned}$$

It follows that

$$\|Lu\|^2 \le \|u\|^2 \left[\int_0^1 \int_0^1 |k(\xi, \xi')|^2 d\xi d\xi' \right]$$

and finally,

$$\|Lu\| \le M\|u\|$$

where M is the bound on the double integral.

■

A linear operator L with domain $\mathcal{D}_L \subset \mathcal{H}$ is *continuous* if given an $\epsilon > 0$, there exists a $\delta > 0$ such that, for every $u_0 \in \mathcal{D}_L$, $\|Lu_0 - Lu\| < \epsilon$, for all $u \in \mathcal{D}_L$ satisfying $\|u_0 - u\| < \delta$. We can interpret this definition to mean that when an operator is continuous, Lu_0 is close to Lu whenever u_0 is close to u. There is an important theorem on interchange of operators and limits that follows immediately from the above definition. A linear operator L with domain $\mathcal{D}_L \subset \mathcal{H}$ is continuous if and only if for every sequence $\{u_n\}_{n=1}^{\infty} \in \mathcal{D}_L$ converging to $u_0 \in \mathcal{D}_L$,

$$Lu_0 = L \lim_{n\to\infty} u_n = \lim_{n\to\infty} Lu_n \tag{1.84}$$

The proof is in two parts. First, we suppose that L is continuous and $\epsilon > 0$ is given. We may select a δ according to the definition of continuity and suppose that $\|u_0 - u_n\| < \delta$. Since u_n is a member of a converging sequence, $\|u_0 - u_n\| < \delta$ for all $n \geq N$. Therefore,

$$\|Lu_0 - Lu_n\| < \epsilon, \qquad n \geq N$$

and

$$Lu_0 = \lim_{n\to\infty} Lu_n$$

This first part of the proof shows that, if an operator is continuous, the operator and the limit can be interchanged. In the second part of the proof, we must show that if the operator and limit can be interchanged, the operator is continuous. This part is not essential to our development and is omitted. The interested reader is referred to [12].

We now give a theorem linking the boundedness and continuity of operators. A linear operator L with domain $\mathcal{D}_L \subset \mathcal{H}$ is bounded if and only if it is continuous. The proof is in two parts. In the first part, we show that if the operator is bounded, it is continuous. Indeed, if L is bounded and $u_0 \in \mathcal{D}_L$,

$$\begin{aligned}\|Lu_0 - Lu\| &= \|L(u_0 - u)\| \\ &\leq \gamma \|u_0 - u\|\end{aligned}$$

for all $u \in \mathcal{D}_L$. Then, given any $\epsilon > 0$, it is easy to find a $\delta > 0$ such that

$$\|Lu_0 - Lu\| < \epsilon$$

whenever

$$\|u_0 - u\| < \delta$$

Indeed, the choice $\delta = \epsilon/\gamma$ is sufficient. In the second part of the proof, we must show that if an operator is continuous, it is bounded. This part is not essential to our development and is omitted. The interested reader is referred to [13].

It is straightforward to show that the differential operator $L = d/d\xi$ is unbounded. The proof is by contradiction. We suppose that $d/d\xi$ is bounded. Then, it is continuous. Therefore, for any $u_n \to u$, we must have

$$Lu = L \lim_{n\to\infty} u_n = \lim_{n\to\infty} Lu_n$$

But, if we choose u_n as a member of the sequence

$$u_n = \frac{\cos n\pi\xi}{n}, \qquad n = 1, 2, \ldots$$

we have

$$\lim_{n\to\infty} u_n = 0$$

and therefore,

$$L \lim_{n\to\infty} u_n = 0$$

But,

$$\lim_{n\to\infty} Lu_n = \lim_{n\to\infty} (-\pi \sin n\pi\xi)$$

and this limit is undefined. We therefore have arrived at a contradiction, and we conclude that $d/d\xi$ is unbounded.

Given the concepts of continuity and boundedness of a linear operator, we can show that a bounded linear operator is uniquely determined by a matrix. Indeed, let $\{z_k\}_{k=1}^{\infty}$ be a basis for $\mathcal{H}$. Let L be a bounded linear operator with

$$Lu = f$$

We expand u in the basis as follows:

$$u = \lim_{n\to\infty} \sum_{k=1}^{n} \alpha_k z_k \tag{1.85}$$

Since boundedness implies continuity,

$$Lu = L \lim_{n\to\infty} \sum_{k=1}^{n} \alpha_k z_k = \lim_{n\to\infty} L \sum_{k=1}^{n} \alpha_k z_k$$

By the linearity of the operator L, we then have

$$\lim_{n\to\infty} \sum_{k=1}^{n} \alpha_k L z_k = f \tag{1.86}$$

We take the inner product of both sides of (1.86) with a member of the basis set to give

$$\langle \lim_{n\to\infty} \sum_{k=1}^{n} \alpha_k L z_k, z_j \rangle = \langle f, z_j \rangle, \qquad j = 1, 2, \ldots$$

By continuity of the inner product and the rules for inner products, we obtain

$$\lim_{n\to\infty} \sum_{k=1}^{n} \alpha_k \langle L z_k, z_j \rangle = \langle f, z_j \rangle, \qquad j = 1, 2, \ldots \tag{1.87}$$

Equation (1.87) is a matrix equation that can be written in standard matrix notation as follows:

$$\begin{bmatrix} \langle Lz_1, z_1 \rangle & \langle Lz_2, z_1 \rangle & \cdots \\ \langle Lz_1, z_2 \rangle & \langle Lz_2, z_2 \rangle & \cdots \\ \vdots & \vdots & \vdots \end{bmatrix} \begin{bmatrix} \alpha_1 \\ \alpha_2 \\ \vdots \end{bmatrix} = \begin{bmatrix} \langle f, z_1 \rangle \\ \langle f, z_2 \rangle \\ \vdots \end{bmatrix} \tag{1.88}$$

If this matrix can be inverted to give the coefficients $\alpha_1, \alpha_2, \ldots$, substitution into (1.85) completes the determination of u.

EXAMPLE 1.26 On the real Hilbert space $\mathcal{L}_2(0, 1)$, consider the integral equation

$$f(\xi) = -\mu u(\xi) + \int_0^1 k(\xi, \xi') u(\xi') d\xi' \tag{1.89}$$

where

$$k(\xi, \xi') = \begin{cases} \xi(1 - \xi'), & 0 \le \xi \le \xi' \\ \xi'(1 - \xi), & \xi' \le \xi \le 1 \end{cases} \tag{1.90}$$

In operator notation,

$$(L - \mu I)u = f, \qquad \xi \in (0, 1) \tag{1.91}$$

where I is the identity operator. The operator $L - \mu I$ is bounded. We leave the proof for Problem 1.26. We wish to obtain the matrix representation and thereby solve the integral equation. We define the inner product for the space as in (1.63). For basis functions, we choose

$$z_n = \sin n\pi\xi, \qquad n = 1, 2, \ldots \tag{1.92}$$

Then, operating on any member of the basis set, we obtain

$$(L - \mu I) z_n = -\mu \sin n\pi\xi + \int_0^1 k(\xi, \xi') \sin n\pi\xi' d\xi' \tag{1.93}$$

But, using (1.90), we find that

$$\int_0^1 k(\xi,\xi')\sin n\pi\xi' d\xi' = (1-\xi)\int_0^{\xi} \xi' \sin n\pi\xi' d\xi' + \xi \int_{\xi}^{1} (1-\xi')\sin n\pi\xi' d\xi'$$

After some elementary integrations, we obtain the general matrix element in the square matrix in (1.88), viz.

$$\begin{aligned}\langle (L-\mu I)z_n, z_m\rangle &= \left[\frac{1}{(n\pi)^2} - \mu\right]\langle z_n, z_m\rangle \\ &= \frac{1}{2}\left[\frac{1}{(n\pi)^2} - \mu\right]\delta_{nm} \qquad (1.94)\end{aligned}$$

where the inner product is the usual inner product for $\mathcal{L}_2(0,1)$ and δ_{nm} has been defined in (1.33). The matrix representation in (1.88) therefore diagonalizes, and the inversion yields

$$\alpha_k = \frac{2}{\frac{1}{(k\pi)^2} - \mu}\langle f, z_k\rangle, \qquad k = 1, 2, \ldots \qquad (1.95)$$

Substitution of (1.95) into (1.85) yields the final result, viz.

$$u(\xi) = 2\sum_{k=1}^{\infty} \frac{\int_0^1 f(\xi')\sin k\pi\xi' d\xi'}{\frac{1}{(k\pi)^2} - \mu}\sin k\pi\xi \qquad (1.96)$$

■

In the above example of representation of an operator by a matrix, the choice of the basis functions resulted in diagonalization of the matrix and, therefore, trivial matrix inversion. There are many operators, however, that do not have properties that result in this diagonalization. These concepts are better understood after a study of operator properties and resulting Green's functions and spectral representations in the next two chapters.

An important collection of operators for which there are established convergence criteria are nonnegative, positive, and positive-definite operators. The reader is cautioned that there is little uniformity of notation concerning these operators in the literature. For the purposes herein, an operator L is *nonnegative* if $\langle Lx, x\rangle \geq 0$, for all $x \in \mathcal{D}_L$. An operator is *positive* if $\langle Lx, x\rangle > 0$, for all $x \neq 0$ in $\mathcal{D}_L$. An operator is *positive-definite* if $\langle Lx, x\rangle \geq c^2\|x\|^2$, for $c > 0$ and $x \in \mathcal{D}_L$. An operator L is *symmetric* if $\langle Lx, x\rangle = \langle x, Lx\rangle$. It is easy to show that nonnegative, positive, and positive-definite operators are symmetric. In fact, any operator having the property that $\langle Lx, x\rangle$ is real is symmetric. Indeed,

$$\langle x, Lx\rangle = \overline{\langle Lx, x\rangle} = \langle Lx, x\rangle$$

A special inner product and norm [17], associated with positive and positive-definite operators, are useful in relating convergence criteria. Define the *energy inner product* with respect to the operator L by

$$[x, y] = \langle Lx, y \rangle \tag{1.97}$$

With this inner product definition, $\mathcal{D}_L$ becomes a Hilbert space $\mathcal{H}_L$. The associated *energy norm* in $\mathcal{H}_L$ is given by

$$\left| x \right| = \sqrt{\langle Lx, x \rangle} \tag{1.98}$$

We emphasize that the operator L must be positive for (1.98) to satisfy the basic definitions of a norm. Indeed, the energy inner product and norm defined in (1.97) and (1.98) must be shown in each case to satisfy the rules for inner products and norms. For positive-definite operators, we can prove the following important relationship between norms:

$$\|x\| \leq \frac{1}{c} \left| x \right| \tag{1.99}$$

Indeed,

$$\left| x \right|^2 = \langle Lx, x \rangle \geq c^2 \|x\|^2$$

Therefore,

$$\|x\|^2 \leq \frac{1}{c^2} \left| x \right|^2$$

Taking the square root of both sides yields the desired result.

Among the many forms of convergence criteria, there are several types that are particularly useful in numerical methods in electromagnetics. For a sequence $\{u_n\} \subset \mathcal{H}$, u_n *converges to* u is written

$$u_n \rightarrow u \tag{1.100}$$

and means that

$$\lim_{n\to\infty} \|u_n - u\| = 0 \tag{1.101}$$

The statement u_n *converges in energy to* u is written

$$u_n \xrightarrow{e} u \tag{1.102}$$

and means that

$$\lim_{n\to\infty} \left| u_n - u \right| = 0 \tag{1.103}$$

The statement u_n *converges weakly to* u is written

$$u_n \xrightarrow{w} u \tag{1.104}$$

and means that for every $g \in \mathcal{H}$

$$\lim_{n\to\infty} |\langle u_n - u, g\rangle| = 0 \tag{1.105}$$

It is straightforward to show the following relationships among the types of convergence:

A. If $\|Lu_n\|$ is bounded, convergence implies convergence in energy.

B. Convergence implies weak convergence.

C. Convergence in energy implies $Lu_n \xrightarrow{w} f$. The weak convergence is for those g, defined by (1.105), in $\mathcal{H}_L$. If, however, $\|Lu_n\|$ is bounded, then $Lu_n \xrightarrow{w} f$ in $\mathcal{H}$.

D. If L is positive-definite, convergence in energy implies convergence.

We first prove Property A. We have

$$\begin{aligned} \left| u - u_n \right|^2 &= |\langle L(u_n - u), u_n - u\rangle| \le \|L(u_n - u)\| \, \|u_n - u\| \\ &= \|Lu_n - Lu\| \, \|u_n - u\| \le (\|Lu_n\| + \|Lu\|) \, \|u_n - u\| \end{aligned}$$

Since, by hypothesis, $\|Lu_n\|$ is bounded and $u_n \to u$, a limiting operation gives

$$\lim_{n\to\infty} \left| u_n - u \right|^2 = 0$$

To prove Property B, we use the CSB inequality to give

$$|\langle u_n - u, g\rangle| \le \|u_n - u\| \, \|g\|$$

for any g in $\mathcal{H}$. Taking the limit yields the desired result, viz.

$$\lim_{n\to\infty} |\langle u_n - u, g\rangle| = 0$$

To prove Property C, we have

$$|\langle Lu_n - f, g\rangle| = |\langle L(u_n - u), g\rangle| = |[u_n - u, g]| \le \left| u_n - u \right| \left| g \right|$$

where we have used the CSB inequality on Hilbert space $\mathcal{H}_L$. By hypothesis, we have convergence in energy. Therefore,

$$\lim_{n\to\infty} |\langle Lu_n - f, g\rangle| = 0$$

for $g \in \mathcal{H}_L$. This procedure proves the first half of Property C. The proof of the second half is based on the Hilbert space $\mathcal{H}_L$ being *dense* in $\mathcal{H}$ and is omitted. (See [17, p. 24–25].) To prove Property D, we write

$$\|u_n - u\| \leq \frac{1}{c} \left| u_n - u \right|$$

Taking the limit as $n \to \infty$ yields the desired result.

1.8 METHOD OF MOMENTS

The purpose of this section is to introduce the Method of Moments in a general way and develop various special cases. Emphasis is on convergence and error minimization.

If L is a linear operator, an approximate solution to $Lu = f$ is given by the following procedure. For L an operator in $\mathcal{H}$, consider

$$Lu - f = 0 \tag{1.106}$$

where $u \in \mathcal{D}_L$, $f \in \mathcal{R}_L$. Define the linearly independent sets $\{\phi_k\}_{k=1}^n \subset \mathcal{D}_L$ and $\{w_k\}_{k=1}^n$, where ϕ_k and w_k are called *expansion* functions and *weighting* functions, respectively. Define a sequence of approximants to u by

$$u_n = \sum_{k=1}^{n} \alpha_k \phi_k, \qquad n = 1, 2, \ldots \tag{1.107}$$

A matrix equation is formed in (1.106) by the condition that, upon replacement of u by u_n, the left side shall be orthogonal to the sequence $\{w_k\}$. We have

$$\langle Lu_n - f, w_m \rangle = 0, \qquad m = 1, 2, \ldots, n \tag{1.108}$$

Substitution of (1.107) into (1.108) and use of (1.25) gives the matrix equation of the *Method of Moments* [18],[19], viz.

$$\sum_{k=1}^{n} \alpha_k \langle L\phi_k, w_m \rangle = \langle f, w_m \rangle, \qquad m = 1, 2, \ldots, n \tag{1.109}$$

Note that the *exact* operator equation (1.106) in a Hilbert space $\mathcal{H}$ has been transformed into an *approximate* operator equation on Hilbert space $\mathbf{C}_n$, viz.

$$Ax = b \tag{1.110}$$

where, in usual matrix form,

$$x = (\alpha_1 \quad \alpha_2 \quad \cdots \quad \alpha_n)^T \tag{1.111}$$

$$b = (\langle f, w_1 \rangle \quad \langle f, w_2 \rangle \quad \cdots \quad \langle f, w_n \rangle)^T \tag{1.112}$$

$$A = [a_{mk}] \tag{1.113}$$

where T denotes *transpose* and a_{mk} are the individual matrix elements, given by

$$a_{mk} = \langle L\phi_k, w_m \rangle \tag{1.114}$$

We note the following interesting result. If the operator L is bounded, if w_k=ϕ_k, and if the sequence ϕ_k, $k = 1, 2, \ldots$ is a basis for the Hilbert space, then taking the limit as $n \to \infty$ in (1.109) reproduces the result in (1.87). We therefore view the Method of Moments given by (1.109) as an extension to the matrix representation result for bounded operators in (1.87). There remains, however, a basic question concerning the convergence of (1.107) when the sequence $\{\alpha_k\}$ is determined by solution of the matrix equation (1.110).

In the special case where the expansion functions are identical to the weighting functions, the result is *Galerkin's Method* [17], viz.

$$\sum_{k=1}^{n} \alpha_k \langle L\phi_k, \phi_m \rangle = \langle f, \phi_m \rangle, \qquad m = 1, 2, \ldots, n \tag{1.115}$$

If nothing is known about the mathematical properties of the operator L other than its linearity, nothing in general can be said concerning the convergence of the approximants u_n to the solution u. Unfortunately, most of the interesting and practical problems in electromagnetics involve operators that are neither positive nor positive-definite. Therefore, most of the large body of solutions to electromagnetic problems by the Method of Moments lack any sort of mathematical convergence criteria.

If, however, the operator L is positive, we may define the Hilbert space $\mathcal{H}_L$ with the norm given by (1.98). We then write (1.115) as follows:

$$\sum_{k=1}^{n} \alpha_k [\phi_k, \phi_m] = [u, \phi_m], \qquad m = 1, 2, \ldots, n \tag{1.116}$$

We further assume that the sequence $\{\phi_k\}$ is complete in $\mathcal{H}_L$. We may show that, under these circumstances, Galerkin's Method results in convergence in energy. Indeed, since the complete sequence $\{\phi_k\}$ defines $\mathcal{H}_L$, we may apply the Gram–Schmidt orthonormalization procedure. Assuming that the ϕ_m are orthonormal in (1.116), we obtain

$$\alpha_k = [u, \phi_k] \tag{1.117}$$

Substitution into (1.107) gives

$$u_n = \sum_{k=1}^{n} [u, \phi_k]\phi_k \tag{1.118}$$

which is the Fourier series expansion in $\mathcal{H}_L$ of u_n with Fourier coefficients given by (1.117). Therefore,

$$\lim_{n\to\infty} \left| u_n - u \right| = 0 \tag{1.119}$$

By Property C, the result in (1.119) implies that $Lu_n \xrightarrow{w} f$. Unfortunately, nothing can be said about the nearness of u_n to u. If, however, L is positive-definite, Property D states that the approximants converge, viz.

$$\lim_{n\to\infty} \|u_n - u\| = 0 \tag{1.120}$$

In the Galerkin procedure, if the operator L is positive and the sequence $\{\phi_k\}$ is complete in $\mathcal{H}_L$, the method is called the *Rayleigh–Ritz* method. For a classical treatment of the Rayleigh–Ritz method, the reader should consult [17],[20].

For the more general operators often encountered in electromagnetics, a positive operator can be produced by the following procedure. Consider

$$Lu = f \tag{1.121}$$

Let the *adjoint operator* L^* be defined by

$$\langle Lu, v\rangle \;=\; \langle u, L^*v\rangle \tag{1.122}$$

for $u \in \mathcal{D}_L$, $v \in \mathcal{D}_{L^*}$. Then, if the adjoint L^* exists, multiplication of both sides of (1.121) by L^* produces

$$L^*Lu = L^*f \tag{1.123}$$

for any $f \in \mathcal{D}_{L*}$. Provided that $Lu = 0$ has none but the trivial solution, it is easy to show that the operator L^*L is positive. Indeed, $\langle L^*Lu, u\rangle = \|Lu\|^2 > 0$, unless $Lu = 0$. But, $Lu = 0$ implies $u = 0$.

The Method of Moments applied to L^*L gives

$$\sum_{k=1}^{n} \alpha_k \langle L^*L\phi_k, w_m\rangle = \langle L^*f, w_m\rangle, \qquad m = 1, 2, \ldots, n \tag{1.124}$$

The Galerkin specialization follows immediately, viz.

$$\sum_{k=1}^{n} \alpha_k \langle L^* L\phi_k, \phi_m\rangle = \langle L^* f, \phi_m\rangle, \qquad m = 1, 2, \ldots, n \tag{1.125}$$

Since L^*L is positive, if the sequence $\{\phi_k\}$ is complete in $\mathcal{D}_{L^*L}$, (1.125) is the Rayleigh–Ritz method and convergence in energy $u_n \xrightarrow{e} u$ is assured, viz.

$$\lim_{n\to\infty} \left| u_n - u \right| = 0 \tag{1.126}$$

where the energy norm is with respect to the operator L^*L. By properties of the adjoint, (1.125) can also be written

$$\sum_{k=1}^{n} \alpha_k \langle L\phi_k, L\phi_m\rangle = \langle f, L\phi_m\rangle, \qquad m = 1, 2, \ldots, n \tag{1.127}$$

which is the result in the *Method of Least Squares*, more usually derived [20] by minimization of

$$\|Lu_n - f\|^2$$

It is easy to show that (1.126) implies that

$$\lim_{n\to\infty} \|Lu_n - f\| = 0 \tag{1.128}$$

so that $Lu_n \to f$. Unless the operator L^*L is positive-definite, nothing can be said concerning the convergence of u_n to u.

A.1 APPENDIX—PROOF OF PROJECTION THEOREM

In this Appendix, we prove the Projection Theorem, considered in Section 1.6. We restate the theorem here for convenience. Let x be any vector in the Hilbert space $\mathcal{H}$, and let $\mathcal{M} \subset \mathcal{H}$ be a *closed* linear manifold. Then, there is a unique vector $y_0 \in \mathcal{M} \subset \mathcal{H}$ closest to x in the sense that

$$\delta = \inf_{y\in\mathcal{M}} \|x - y\| = \|x - y_0\| \tag{A.1}$$

where *inf* is the *greatest lower bound*, or *infimum*. In other words, y_0 is closest to x in the sense that $\|x - y_0\| \leq \|x - y\|$ for all y in $\mathcal{M}$. Furthermore, the necessary and sufficient condition that y_0 is the unique minimizing vector is that $e = x - y_0$ is in $\mathcal{M}^\perp$. The vector y_0 is called the *projection* of x onto $\mathcal{M}$. The vector e is called the projection of x onto $\mathcal{M}^\perp$.

In proving the Projection Theorem, we begin by noting that the first equality in (A.1) makes sense. Indeed, $\|x - y\|$ is bounded below by zero, and therefore has a greatest lower bound. We next show that there exists at least one vector y_0 closest to x. We begin by asserting that there exists a vector $y_n \in \mathcal{M}$ such that by the definition of infimum,

$$\delta \leq \|x - y_n\| < \delta + \frac{1}{n} \tag{A.2}$$

Taking the limit as $n \to \infty$, we find that

$$\lim_{n\to\infty} \|x - y_n\| = \delta \tag{A.3}$$

Therefore, we can always define a sequence $\{y_n\} \in \mathcal{M}$ such that $\|x - y_n\|$ converges to δ. In (1.36), if we replace x by $x - y_n$ and y by $x - y_m$, we obtain [21], after some rearrangement,

$$\|y_n - y_m\|^2 = 2\|x - y_n\|^2 + 2\|x - y_m\|^2 - 4\|x - \frac{1}{2}(y_n + y_m)\|^2 \tag{A.4}$$

Since $\mathcal{M}$ is a linear manifold, $(y_n + y_m)/2 \in \mathcal{M}$, and we may assert that

$$\|x - \frac{1}{2}(y_n + y_m)\| \geq \delta$$

Therefore,

$$\|y_n - y_m\|^2 \leq 2\|x - y_n\|^2 + 2\|x - y_m\|^2 - 4\delta^2 \tag{A.5}$$

In the limit as $m, n \to \infty$, the right side goes to $2\delta^2 + 2\delta^2 - 4\delta^2 = 0$, and we conclude that the sequence $\{y_n\}$ is Cauchy. Since $\mathcal{H}$ is a Hilbert space and $\mathcal{M}$ is closed, $\mathcal{M}$ is a Hilbert space and Cauchy convergence implies convergence. Therefore, $y_n \to y_0 \in \mathcal{M}$.

We next show that y_0 is unique [22]. Suppose it is not unique. Then, we must have at least two solutions y_0 and $\hat{y}_0$ satisfying $\|x - y_0\| = \|x - \hat{y}_0\| = \delta$. Then,

$$\begin{aligned} \|y_0 - \hat{y}_0\|^2 &= 2\|x - y_0\|^2 + 2\|x - \hat{y}_0\|^2 - 4\|x - \frac{1}{2}(y_0 + \hat{y}_0)\|^2 \\ &\leq 2\delta^2 + 2\delta^2 - 4\delta^2 = 0 \end{aligned} \tag{A.6}$$

Therefore, $y_0 = \hat{y}_0$.

Finally, we show that $e = x - y_0 \in \mathcal{M}^\perp$. We must show that e is orthogonal to every vector in $\mathcal{M}$. Suppose that there exists a vector $z \in \mathcal{M}$ that is not orthogonal to e. Then, we would have [23]

$$\langle e, z\rangle = \langle x - y_0, z\rangle = A \neq 0, \qquad z \in \mathcal{M} \tag{A.7}$$

We define a vector $z_0 \in \mathcal{M}$ such that

$$z_0 = y_0 + \frac{A}{\|z\|^2} z \tag{A.8}$$

Then,

$$\begin{aligned} \|x - z_0\|^2 &= \langle x - y_0 - \frac{A}{\|z\|^2} z, x - y_0 - \frac{A}{\|z\|^2} z\rangle \\ &= \|x - y_0\|^2 - \frac{\overline{A}}{\|z\|^2}\langle x - y_0, z\rangle - \frac{A}{\|z\|^2}\langle z, x - y_0\rangle + \frac{|A|^2}{\|z\|^2} \\ &= \|x - y_0\|^2 - \frac{|A|^2}{\|z\|^2} \end{aligned} \tag{A.9}$$

Therefore,

$$\|x - z_0\| < \|x - y_0\| \tag{A.10}$$

which, by (A.1), is impossible.

PROBLEMS

1.1. Using the rules defining a linear space, show that $0a = \mathbf{0}$ and $-1a = -a$.

1.2. Show that $\mathbf{R}_n$ is a linear space.

1.3. Show that $\mathcal{C}(0,1)$ is a linear space.

1.4. As an extension to Example 1.4, in $\mathbf{R}_2$, let $x_1 = (1, 3)$, $x_2 = (2, 6.00000001)$. Show that x_1 and x_2 are linearly independent. Comment on what might occur in solving this problem on a computer with eight-digit accuracy. (This problem is indicative of the difficulties that can arise in establishing linear independence in numerical experiments in finite length arithmetic.)

1.5. If $x_1, x_2, \ldots, x_n$ is a linearly dependent set, show that at least one member of the set can be written as a linear combination of the other members.

1.6. Show that if $\mathbf{0}$ is a member of the set $x_1, x_2, \ldots, x_n$, the set is linearly dependent.

1.7. Show that if a set containing n vectors is linearly dependent, and if m additional vectors are added to the set, the resulting set of $n+m$ vectors is linearly dependent.

1.8. Show that in $\mathbf{R}_n$ the set of vectors

$$e_1 = (1, 0, \ldots, 0), e_2 = (0, 1, \ldots, 0), \ldots, e_n = (0, 0, \ldots, 1)$$

is linearly independent. Is the same conclusion valid in $\mathbf{C}_n$? Is the set of vectors a basis for $\mathbf{C}_n$?

1.9. Given the basic definition of an inner product space, show that

$$\langle \sum_{k=1}^{n} \alpha_k x_k, y \rangle = \sum_{k=1}^{n} \alpha_k \langle x_k, y \rangle$$

1.10. Show that $C(\alpha, \beta)$ with inner product defined by (1.27) is a real inner product space.

1.11. Consider the linear space of real continuous twice differentiable functions over the interval (0, 1). As a candidate for an inner product for the space, consider

$$\langle f, g \rangle = \int_0^1 f''(\tau) g''(\tau) d\tau$$

where f and g are members of the space and "primes" indicate differentiation. Determine whether $\langle f, g \rangle$ is a legitimate inner product.

1.12. Prove the following corollary to the CSB inequality in (1.35):

$$|\langle x, y \rangle| = \|x\| \, \|y\|$$

if and only if x and y are linearly dependent.

1.13. Prove the following identity:

$$| \, \|x\| - \|y\| \, | \leq \|x - y\|$$

1.14. Consider a complex inner product space with norm induced by the inner product. If x and y are members of the space, prove that

$$\langle x, y \rangle - \langle y, x \rangle = \frac{i}{2} \left(\|x + iy\|^2 - \|x - iy\|^2 \right)$$

and

$$\langle x, y \rangle + \langle y, x \rangle = \frac{1}{2} \left(\|x + y\|^2 - \|x - y\|^2 \right)$$

1.15. Given the following sequence in the space of rational numbers:

$$x_n = \sum_{k=1}^{n} \frac{1}{(k-1)!}$$

First, show that the sequence is Cauchy; next, show that the sequence does not converge in the space. (Indeed, it converges to e; the details can be found in [4].)

1.16. Show that if $\mathcal{S}$ is a linear space, a linear manifold $\mathcal{M} \subset \mathcal{S}$ is also a linear space.

1.17. Let x_k, $k = 1, 2, \ldots, n$ be a linearly independent sequence of vectors in the Hilbert space $\mathcal{H}$. Define $\mathcal{M}$ to be the set of all linear combinations of the n vectors. Prove that $\mathcal{M}$ is a linear manifold.

1.18. Let $\mathbf{R}_\infty$ be the space of all vectors consisting of a countably infinite set of real numbers (components), viz.

$$a = (\alpha_1, \alpha_2, \ldots)$$

where $\alpha_k \in \mathbf{R}$. Let $\mathcal{M}$ be the set of vectors in $\mathbf{R}_\infty$ with only a finite number of the countably infinite number of components different from zero. Show that $\mathcal{M}$ is a linear manifold. Show that $\mathcal{M}$ is not closed. *Hint*: Consider the concept of *closed* as applied specifically to the sequence of vectors

$$x_n = \left(1, \frac{1}{2}, \frac{1}{3}, \ldots, \frac{1}{n}, 0, 0, \ldots\right)$$

1.19. The Legendre functions $P_n(\xi)$, $n = 0, 1, 2, \ldots$, are orthogonal on $\xi \in (-1, 1)$, but they are *not* orthonormal. Create a sequence of *orthonormalized* Legendre functions $\hat{P}_n(\xi)$, $n = 0, 1, 2, \ldots$.

1.20. Given that in $\mathbf{R}_3$, $x_1 = (1, 2, 0)$, $x_2 = (0, 1, 2)$, $x_3 = (1, 0, 1)$.

(a) Prove that $\{x_1, x_2, x_3\}$ is a linearly independent set of vectors.

(b) From the linearly independent set, produce *the first two members* of the associate orthonormal set using the Gram–Schmidt procedure.

1.21. Show that the determinant of the Gram matrix in (1.69) is nonzero if and only if the sequence of vectors $\{y_k\}_{k=1}^M$ is linearly independent [11].

1.22. Let $\mathbf{R}_\infty$ be the space described in Problem 1.18. If $b = (\beta_1, \beta_2, \ldots)$, define an inner product for the space by

$$\langle a, b\rangle = \sum_{k=1}^{\infty} \alpha_k \beta_k$$

Let the norm for the space be induced by the inner product. We restrict $\mathbf{R}_\infty$ to those vectors with finite norm. Define the operator A in $\mathbf{R}_\infty$ by

$$Aa = \left(\alpha_1, \frac{1}{2}\alpha_2, \frac{1}{3}\alpha_3, \ldots\right)$$

Test the operator A for boundedness.

1.23. On the Hilbert space $\mathcal{L}_2(-\alpha, \alpha)$, with inner product

$$\langle f, g\rangle = \int_{-\alpha}^{\alpha} f(\xi)g(\xi)\frac{d\xi}{\sqrt{\alpha^2 - \xi^2}}$$

consider the following integral equation:

$$\int_{-\alpha}^{\alpha} u(\xi) \ln|\eta - \xi| \frac{d\xi}{\sqrt{\alpha^2 - \xi^2}} = f(\eta)$$

This integral equation occurs in diffraction by a slit in a perfectly conducting screen. It can be shown [14] that the operator

$$L = \int_{-\alpha}^{\alpha} (\,\cdot\,) \ln|\eta - \xi| \frac{d\xi}{\sqrt{\alpha^2 - \xi^2}}$$

is bounded. Solve the integral equation by using the Chebyshev polynomials $T_n(\xi/\alpha)$ as a basis for $\mathcal{L}_2(-\alpha, \alpha)$ and obtaining the matrix representation for L. *Hint*: The following are useful integral relations [15]:

$$\int_{-\alpha}^{\alpha} \frac{T_n(\eta/\alpha) \ln|\xi - \eta| d\eta}{\sqrt{\alpha^2 - \eta^2}} = \begin{cases} -\pi \ln(2/\alpha) T_0(\xi/\alpha), & n = 0 \\ -\dfrac{\pi}{n} T_n(\xi/\alpha), & n > 0 \end{cases}$$

$$\int_{-1}^{1} T_m(\xi) T_n(\xi) \frac{d\xi}{\sqrt{1 - \xi^2}} = \begin{cases} 0, & m \neq n \\ \frac{\pi}{2}, & m = n \neq 0 \\ \pi, & m = n = 0 \end{cases}$$

1.24. Let $L = d/d\xi$, and consider the sequence of partial sums

$$u_n = \sum_{k=1}^{n} \frac{1}{k} \cos k\pi\xi$$

It is well-known [16] that

$$\lim_{n\to\infty} u_n = -\ln\left(2 \sin \frac{\pi\xi}{2}\right)$$

Show that $\lim_{n\to\infty} Lu_n$ is undefined. The problem is that L is unbounded. This result is an example of the fact that a Fourier series cannot always be differentiated term by term.

1.25. Let $L = d/d\xi$, and consider the sequence of partial sums

$$u_n = \sum_{k=1}^{n} \frac{1}{k^2} \cos k\pi\xi$$

Using well-known series summation results [16], show that, although L is unbounded, in this case the operator and limit *can* be interchanged.

1.26. Show that the operator in (1.91) with kernel defined in (1.90) is bounded.

1.27. Consider Hilbert space $\mathcal{L}_2(-1, 1)$ with inner product

$$\langle u, v\rangle = \int_{-1}^{1} u(\xi)v(\xi)d\xi$$

where all functions are real-valued. Consider the following function $f(\xi)$:

$$f(\xi) = \xi - 4\xi^3$$

Construct a function $g(\xi)$ orthogonal to $f(\xi)$. Adjust $g(\xi)$ such that it has unit norm

$$\|g(\xi)\| = 1$$

1.28. Given that, in $\mathbf{R}_3$, $x_1 = (1, 1, 0)$, $x_2 = (0, 1, 1)$, $x_3 = (1, 0, 1)$.

(a) Prove that $\{x_1, x_2, x_3\}$ is a linearly independent set of vectors.

(b) Using the Gram–Schmidt procedure, produce the first two members $\{e_1, e_2\}$ of the orthonormal set obtained from this linearly independent set.

1.29. Given Hilbert space $\mathcal{L}_2(0, 2\pi)$ with inner product

$$\langle u, v\rangle = \int_{0}^{2\pi} u(\xi)\overline{v(\xi)}d\xi$$

where members of the space are complex functions.

(a) Show that the sequence

$$\{e^{in\xi}\}_{n=-\infty}^{\infty}$$

is an orthogonal sequence.

(b) Produce an orthonormal (O.N.) sequence from the orthogonal sequence.

(c) Using the members of the O.N. sequence contained on $-N \leq n \leq N$, where N is a positive integer, find the best approximation to

$$f(\xi) = \sin\left(\frac{\xi - 2a}{2}\right), \qquad a \in \mathbf{R}$$

in the sense given in Section 1.6, *Best Approximation*.

1.30. Consider the real Hilbert space $\mathcal{L}_2(-1, 1)$. For $f(\xi), g(\xi) \in \mathcal{L}_2(-1, 1)$, define an inner product

$$\langle f, g\rangle = \int_{-1}^{1} f(\xi)g(\xi)\frac{d\xi}{\sqrt{1 - \xi^2}}$$

Determine whether this definition results in a legitimate inner product.

1.31. Consider the real Hilbert space $\mathcal{L}_2(0, 1)$ with inner product

$$\langle f, g \rangle = \int_0^1 f(\xi) g(\xi) d\xi$$

where $f(\xi), g(\xi) \in \mathcal{L}_2(0, 1)$. Suppose that

$$f(\xi) = 1 - \frac{\xi}{2}$$

(a) By the method of best approximation, approximate $f(\xi)$ by $\hat{f}(\xi)$, where $\hat{f}(\xi)$ is a linear combination constructed from the orthonormal set $\{\sqrt{\epsilon_k} \cos k\pi\xi\}_{k=0}^{1}$ in $\mathcal{L}_2(0, 1)$. In the orthonormal set, ϵ_k is 1 for $k = 0$ and 2 for $k \neq 0$.

(b) Calculate the norm of the error $\| f(\xi) - \hat{f}(\xi) \|$.

1.32. Consider Euclidean space $\mathbf{R}_4$. Define vectors a and b in $\mathbf{R}_4$ by

$$a = (\alpha_1, \cdots, \alpha_4)$$
$$b = (\beta_1, \cdots, \beta_4)$$

Define an inner product for the space by

$$\langle a, b \rangle = \sum_{k=1}^{4} \alpha_k \beta_k$$

Consider those vectors a in $\mathbf{R}_4$ restricted by

$$\alpha_1 - \alpha_2 = \alpha_3 - \alpha_4 = 0$$

(a) Show that all vectors with this restriction form a linear manifold $\mathcal{M}$.

(b) Find all vectors b that are members of $\mathcal{M}^{\perp}$.

REFERENCES

[1] Stewart, G.W. (1973), *Introduction to Matrix Computations.* New York: Academic Press, 54–56.

[2] Hardy, G.H. (1967), *A Course in Pure Mathematics.* London: Cambridge University Press, 1–2.

[3] De Lillo, N.J. (1982), *Advanced Calculus with Applications.* New York: Macmillan, 32.

[4] Stakgold, I. (1967), *Boundary Value Problems of Mathematical Physics, Vol. 1.* New York: Macmillan, 101–102.

[5] Mac Duffee, C.C. (1940), *Introduction to Abstract Algebra.* New York: Wiley, chapter VI.

[6] Greenberg, M.D. (1978), *Foundations of Applied Mathematics.* Englewood Cliffs, NJ: Prentice-Hall, 319–320.

[7] Helmberg, G. (1975), *Introduction to Spectral Theory in Hilbert Space.* Amsterdam: North-Holland, 20–21.

[8] Ibid., Appendix B.

[9] Shilov, G.E. (1961), *An Introduction to the Theory of Linear Spaces.* Englewood Cliffs, NJ: Prentice-Hall, sect. 83–85.

[10] Stakgold, I. (1979), *Green's Functions and Boundary Value Problems.* New York: Wiley-Interscience, 36–41.

[11] Luenberger, D.G. (1969), *Optimization by Vector Space Methods.* New York: Wiley, 56–57.

[12] op.cit. Helmberg, 73.

[13] Akhiezer, N.I., and I.M. Glazman (1961), *Theory of Linear Operators in Hilbert Space.* New York: Frederick Ungar, 33,39.

[14] Dudley, D.G. (1985), Error minimization and convergence in numerical methods, *Electromagnetics* 5: 89–97.

[15] Butler, C.M., and D.R. Wilton (1980), General analysis of narrow strips and slots, *IEEE Trans. Antennas Propagat.* AP-28: 42–48.

[16] Abramowitz, M., and I.A. Stegun (Eds.) (1964), *Handbook of Mathematical Functions*, National Bureau of Standards, Applied Mathematics Series, 55, Superintendent of Documents, U.S. Government Printing Office, Washington, DC 20402, 1005.

[17] Mikhlin, S.G. (1965), *The Problem of the Minimum of a Quadratic Functional.* San Francisco: Holden-Day.

[18] Harrington, R.F. (1968), *Field Computation by Moment Methods.* New York: Macmillan.

[19] Harrington, R.F. (1967), "Matrix methods for field problems," *Proc. IEEE* 55: 136–149.

[20] op.cit. Stakgold (1967), vol. 2, sect. 8.10.

[21] Krall, A.M. (1973), *Linear Methods of Applied Analysis.* Reading, MA: Addison-Wesley, 181–182.

[22] Ibid., 182.

[23] op.cit. Akhiezer and Glazman, 10.

2

The Green's Function Method

2.1 INTRODUCTION

In this chapter, we begin our study of linear ordinary differential equations of second order. Our goal is to develop a procedure whereby we can solve the differential equations using fundamental solutions called *Green's functions*.

We begin with a brief discussion of the delta function. We follow with a description of the Sturm–Liouville operator L and its properties. We define three types of Sturm–Liouville problems and investigate their properties. In all three types, we examine the role of the operator L and its *adjoint operator* L^*. These operators are used to define the Green's function and the adjoint Green's function, respectively. Our study culminates in a procedure for applying the Green's function and/or the adjoint Green's function in solving the differential equation $Lu = f$.

2.2 DELTA FUNCTION

The concept of the delta function arises when we wish to fix attention on the value of a function $f(x)$ at a given point x_0. Mathematically, we seek an operator T that transforms a function $f(x)$, continuous at x_0, into $f(x_0)$, the value of the function at x_0. In equation form, we require T such that

$$T[f(x)] = f(x_0) \tag{2.1}$$

We begin by considering the *pulse function* $p_\epsilon(x - x_0)$, defined by

$$p_\epsilon(x - x_0) = \begin{cases} \frac{1}{2\epsilon}, & x_0 - \epsilon < x < x_0 + \epsilon \\ 0, & \text{otherwise} \end{cases} \tag{2.2}$$

Note that, regardless of the value of ϵ, the area under the pulse is unity. Indeed, if (a, b) is any interval containing $(x_0 - \epsilon, x_0 + \epsilon)$,

$$\int_a^b p_\epsilon(x - x_0)dx = \int_{x_0-\epsilon}^{x_0+\epsilon} \frac{1}{2\epsilon}\, dx = 1 \tag{2.3}$$

An important property of the pulse function is that it is even about x_0, viz.

$$p_\epsilon(x - x_0) = p_\epsilon(x_0 - x) \tag{2.4}$$

This property can be proved by interchanging x and x_0 in (2.2). The details are left for the problems. Multiplying the pulse function by $f(x)$ and integrating over any interval containing the pulse gives (Fig. 2-1)

$$\int_a^b f(x)p_\epsilon(x - x_0)dx = \frac{1}{2\epsilon}\int_{x_0-\epsilon}^{x_0+\epsilon} f(x)dx \tag{2.5}$$

By the mean value theorem for integrals [1], if $\hat{f}$ is the mean value of $f(x)$ on the interval $x \in (x_0 - \epsilon, x_0 + \epsilon)$,

$$\int_a^b f(x)p_\epsilon(x - x_0)dx = \hat{f} \tag{2.6}$$

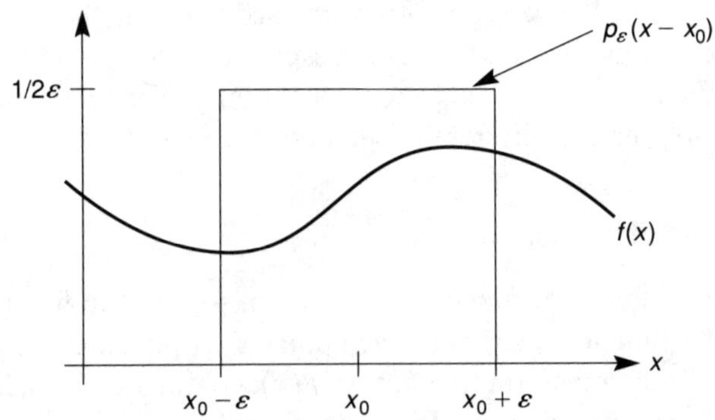

Fig. 2-1 Pulse function $p_\epsilon(x - x_0)$ and function $f(x)$.

Taking the limit as $\epsilon \to 0$, we have

$$\lim_{\epsilon\to 0}\int_a^b f(x)p_\epsilon(x-x_0)dx = f(x_0), \qquad x_0 \in (a,b) \tag{2.7}$$

The integration followed by the limiting operation in (2.7) transforms $f(x)$ to $f(x_0)$, the value of the function at x_0.

EXAMPLE 2.1 Let $f(x) = x^2$, $x_0 = 0$, and $x_0 \in (a,b)$. In this case, $f(x)$ is continuous at $x = 0$ and we have

$$\lim_{\epsilon\to 0}\int_a^b f(x)p_\epsilon(x-x_0)dx = \lim_{\epsilon\to 0}\left[\frac{1}{2\epsilon}\int_{-\epsilon}^{\epsilon} x^2 dx\right]$$
$$= \lim_{\epsilon\to 0}\left(\frac{\epsilon^2}{3}\right) = 0$$

Since $f(0) = 0$, we have verified (2.7).

■

EXAMPLE 2.2 Let $f(x) = \cos x$, $x_0 = \pi/3$, and $x_0 \in (a,b)$. In this case, $f(x)$ is continuous at $x = \pi/3$ and we have

$$\lim_{\epsilon\to 0}\int_a^b f(x)p_\epsilon(x-x_0)dx = \lim_{\epsilon\to 0}\frac{1}{2\epsilon}\int_{\frac{\pi}{3}-\epsilon}^{\frac{\pi}{3}+\epsilon} \cos x\, dx$$
$$= \lim_{\epsilon\to 0}\left\{\frac{1}{2\epsilon}\left[\sin\left(\frac{\pi}{3}+\epsilon\right) - \sin\left(\frac{\pi}{3}-\epsilon\right)\right]\right\} = \frac{1}{2}$$

Since $f(\pi/3) = 1/2$, we have again verified (2.7).

■

Expression (2.7) forms the cornerstone of our definition of the delta function, as follows:

$$\int_a^b f(x)\delta(x-x_0)dx = \lim_{\epsilon\to 0}\int_a^b f(x)p_\epsilon(x-x_0)dx \tag{2.8}$$

so that

$$\int_a^b f(x)\delta(x-x_0)dx = f(x_0) \tag{2.9}$$

for any x_0 in the interval (a,b). Note in (2.2) that as ϵ becomes smaller, the pulse function becomes narrower and higher while maintaining unit area. If the limit in (2.7) could be taken under the integral, we would have

$$\delta(x-x_0) \stackrel{s}{=} \lim_{\epsilon\to 0}[p_\epsilon(x-x_0)] \tag{2.10}$$

Since this limit does not exist, the interchange of limit and integration in (2.8) is not valid. We have therefore placed an "s" over the equality in (2.10) to indicate *symbolic equality* only.

The delta function $\delta(x - x_0)$ has two remarkable properties. Symbolically, it is a function that is zero everywhere except at $x = x_0$, where it is undefined. Second, when integrated against a function f that is continuous at x_0, it yields the value of the function at x_0. We note that (2.9) defines the operator T we were seeking in (2.1). Indeed, comparing (2.1) and (2.9) yields

$$T = \int_a^b (\bullet)\delta(x - x_0)dx \tag{2.11}$$

where $(\bullet)$ indicates the position of the function upon which T operates.

From the basic definition of the delta function in (2.9), we obtain some additional relations. If we set x_0 equal to zero, we find

$$\int_a^b f(x)\delta(x)dx = f(0) \tag{2.12}$$

Also, if in (2.9) we set $f(x) = 1$, we obtain

$$\int_a^b \delta(x - x_0)dx = 1 \tag{2.13}$$

Finally, from (2.4) and (2.8), we conclude symbolically that

$$\delta(x - x_0) \overset{s}{=} \delta(x_0 - x) \tag{2.14}$$

In concluding our development of the delta function and its properties, we remark that there are certain difficulties with the definitions. Indeed, any function that is zero everywhere except at one point must produce zero when Riemann integrated over any interval containing the point. The result in (2.13), for example, is therefore unacceptable in the Riemann sense. To interpret the integral, it seems that we must return to the basic definition in (2.8). The mathematical acceptability of integrals involving the delta function have, however, been formalized in the Theory of Distributions, introduced by Schwartz [2]. In the theory, the delta function is called a *generalized function*, and the integral in (2.9) is said to exist in the *distributional sense*. Although the theory is beyond the scope of this book, the interested reader can find introductory treatments in [3],[4].

The central role played by the delta function in the solution to certain differential equations becomes apparent in the following argument.

Suppose we wish to solve the equation

$$Lu = f \tag{2.15}$$

where L is a differential operator. Formally, the solution is given by multiplying both sides of (2.15) by the inverse operator, viz.

$$L^{-1}Lu = L^{-1}f$$

or

$$u = L^{-1}f \tag{2.16}$$

Since L is a differential operator, we shall assume that its inverse is an integral operator with kernel $g(x, \xi)$, so that

$$u(x) = \int g(x, \xi)f(\xi)d\xi \tag{2.17}$$

Substitution into (2.15) gives

$$\begin{aligned} f(x) &= L[u(x)] \\ &= L\int g(x, \xi)f(\xi)d\xi \\ &= \int Lg(x, \xi)f(\xi)d\xi \end{aligned} \tag{2.18}$$

where we have assumed, without proof, that we can move the operator L inside the integral. But, from the properties of the delta function, we have

$$f(x) = \int_a^b \delta(x - \xi)f(\xi)d\xi \tag{2.19}$$

for $x \in (a, b)$. Comparing (2.18) and (2.19), we identify

$$Lg(x, \xi) = \delta(x - \xi) \tag{2.20}$$

Presumably, if we can solve (2.20), then the solution to (2.15) is given explicitly by (2.17). The kernel $g(x, \xi)$ is called the *Green's function* for the problem.

It is the purpose of this chapter to formalize and structure the introductory ideas above. The result will be the solution to a class of linear ordinary differential equations of second order by the *Green's function method.*

2.3 STURM–LIOUVILLE OPERATOR THEORY

Consider the following linear, ordinary, differential equation of second order:

$$a_0(x)\frac{d^2u(x)}{dx^2}+a_1(x)\frac{du(x)}{dx}+a_2(x)u(x)-\lambda u(x) = f(x), \qquad a < x < b \tag{2.21}$$

where λ is, in general, a complex parameter independent of x. The functions a_0, a_1, and a_2 are real and assumed to have the following properties [5],[6]:

a. a_2, da_1/dx, and d^2a_0/dx^2 are continuous in $a \le x \le b$

b. $a_0 \neq 0$ in $a < x < b$

In (2.21), we also require that $u(x)$ be twice differentiable and that $f(x)$ be piecewise continuous. We may always recast this differential equation in *Sturm–Liouville form*, as follows:

$$-\frac{1}{w(x)}\frac{d}{dx}\left[p(x)\frac{du(x)}{dx}\right] + q(x)u(x) - \lambda u(x) = f(x) \tag{2.22}$$

The necessary coefficient transformations are given by [7]

$$q(x) = a_2(x) \tag{2.23}$$

$$p(x) = \exp\left[\int^x \frac{a_1(t)}{a_0(t)}dt\right] \tag{2.24}$$

$$w(x) = -\frac{p(x)}{a_0(x)} \tag{2.25}$$

We may verify these transformations by substituting (2.23)–(2.25) into (2.22) to produce (2.21). The details are left for the problems. We rewrite (2.22) in operator notation as follows:

$$(L - \lambda)u = f \tag{2.26}$$

where we identify the *Sturm–Liouville operator* L, viz.

$$L = -\frac{1}{w(x)}\frac{d}{dx}\left[p(x)\frac{d}{dx}\right] + q(x) \tag{2.27}$$

For the remainder of this chapter, without loss of generality, L will always mean the Sturm–Liouville operator in (2.27).

EXAMPLE 2.3 Consider Bessel's equation of order ν, given by

$$-u'' - \frac{1}{x}u' + \left(\frac{\nu^2}{x^2} - k^2\right)u = f \tag{2.28}$$

where "prime" indicates differentiation with respect to x. Comparing to (2.21), we identify

$$\begin{aligned}\lambda &= k^2\\ a_0 &= -1\\ a_1 &= -(1/x)\\ a_2 &= (\nu/x)^2\end{aligned}$$

To transform to Sturm–Liouville form, we use (2.23)–(2.25) and obtain

$$\begin{aligned}q(x) &= \frac{\nu^2}{x^2}\\ p(x) &= x\\ w(x) &= x\end{aligned}$$

so that

$$-\frac{1}{x}(xu')' + \left(\frac{\nu^2}{x^2} - k^2\right)u = f \tag{2.29}$$

■

EXAMPLE 2.4 Consider Bessel's equation, given by (2.28), in a slightly different form, viz.

$$-x^2u'' - xu' - \left[(kx)^2 - \nu^2\right]u = \hat{f} \tag{2.30}$$

We note that (2.30) is obtained simply by multiplying both sides of (2.28) by x^2. In this case, we identify

$$\begin{aligned}\lambda &= -\nu^2\\ a_0 &= -x^2\\ a_1 &= -x\\ a_2 &= -(kx)^2\end{aligned}$$

Using (2.23)–(2.25), we obtain

$$\begin{aligned}q(x) &= -(kx)^2\\ p(x) &= x\\ w(x) &= \frac{1}{x}\end{aligned}$$

so that in Sturm–Liouville form,

$$-x(xu')' - (kx)^2 u + \nu^2 u = \hat{f} \tag{2.31}$$

We note, in particular, that the weighting function $w(x)$ differs from that in Example 2.3.

■

It might appear that the distinction between (2.29) and (2.31) is trivial since the latter can be obtained from the former by dividing by x^2. However, the difference in the weighting functions between (2.29) and (2.31) changes the Hilbert space, and makes a major difference in spectral representations associated with the radial portion of the Helmholtz equation in cylindrical coordinates [8], as we shall find in Chapter 4.

EXAMPLE 2.5 Consider Legendre's equation on the interval $x \in (-1, 1)$, as follows:

$$-(1 - x^2)u'' + 2xu' - n(n+1)u = f \tag{2.32}$$

We identify

$$\begin{aligned} \lambda &= n(n+1) \\ a_0 &= -(1 - x^2) \\ a_1 &= 2x \\ a_2 &= 0 \end{aligned}$$

Using (2.23)–(2.25), we obtain

$$\begin{aligned} q(x) &= 0 \\ p(x) &= 1 - x^2 \\ w(x) &= 1 \end{aligned}$$

so that in Sturm–Liouville form,

$$-\left[(1 - x^2)u'\right]' - n(n+1)u = f \tag{2.33}$$

■

The Sturm–Liouville form of the second order differential equation, given by (2.22), plays a central role in the solution of electromagnetic boundary value problems. We distinguish three forms of the Sturm–Liouville problem, which we consider in the next three sections.

2.4 STURM–LIOUVILLE PROBLEM OF THE FIRST KIND

For the first form of the Sturm–Liouville problem, we consider $(L-\lambda)u = f$ over a finite interval $x \in (a, b)$ and for real λ and real f. For $-\infty < a < b < \infty$, consider the Hilbert space $\mathcal{L}_2(a, b)$ with real inner product

$$\langle u, v\rangle = \int_a^b u(x)v(x)w(x)dx \tag{2.34}$$

for all $u, v \in \mathcal{L}_2(a, b)$. We define the *Sturm–Liouville Problem of the First Kind*, abbreviated SLP1, as follows:

$$L_\lambda u = f, \qquad a < x < b \tag{2.35}$$

where

$$L_\lambda = L - \lambda \tag{2.36}$$

and where

$$L = -\frac{1}{w(x)}\frac{d}{dx}\left[p(x)\frac{d}{dx}\right] + q(x) \tag{2.37}$$

We impose the following restrictions [9]:

a. p, p', q, w are real and continuous for $a \le x \le b$
b. $p(x) > 0$, $w(x) > 0$ for $a \le x \le b$
c. λ is real and independent of x

In addition, we require $u(x) \in \mathcal{D}_L \subset \mathcal{L}_2(a, b)$, where $\mathcal{D}_L$ is the domain of the operator L. Because we are dealing with second-order differential operators, the domain is restricted to those functions that are twice differentiable. Finally, we require that $u(x)$ satisfy two boundary conditions as follows:

$$B_1(u) = \alpha = \alpha_{11}u(a) + \alpha_{12}u'(a) + \alpha_{13}u(b) + \alpha_{14}u'(b) \tag{2.38}$$

$$B_2(u) = \beta = \alpha_{21}u(a) + \alpha_{22}u'(a) + \alpha_{23}u(b) + \alpha_{24}u'(b) \tag{2.39}$$

where, for SLP1, α, β, and α_{ij} are real. Typically, in (2.38), if α is nonzero, the boundary condition is said to be *inhomogeneous*. If $\alpha = 0$, the boundary condition is *homogeneous*.

There are several important special cases contained in the boundary conditions in (2.38) and (2.39). A boundary condition is *unmixed* if it involves conditions on $u(x)$ at one boundary only. If SLP1 involves an unmixed condition at one end of the boundary and an unmixed condition at the other end, we refer to this case as SLP1 with *unmixed conditions*. The most general case of unmixed boundary conditions is $\alpha_{13} = \alpha_{14} = \alpha_{21} = \alpha_{22} = 0$, so that

$$B_1(u) = \alpha = \alpha_{11}u(a) + \alpha_{12}u'(a) \tag{2.40}$$

$$B_2(u) = \beta = \alpha_{23}u(b) + \alpha_{24}u'(b) \tag{2.41}$$

The relations in (2.38) and (2.39) are said to be *initial* conditions if $\alpha_{11} = \alpha_{22} = 1$ and all other α_{ij} coefficients are zero, so that

$$B_1(u) = \alpha = u(a) \tag{2.42}$$

$$B_2(u) = \beta = u'(a) \tag{2.43}$$

The two conditions in (2.38) and (2.39) are *periodic* if the value of the function $u(x)$ at one boundary is identical to the value at the other boundary, and if the value of the derivative $u'(x)$ at one boundary is identical to the value at the other boundary. To produce the periodic conditions, we require $\alpha_{11} = -\alpha_{13} = \alpha_{22} = -\alpha_{24} = 1$ and all other coefficients zero, so that

$$u(a) = u(b) \tag{2.44}$$

$$u'(a) = u'(b) \tag{2.45}$$

EXAMPLE 2.6 Consider the following differential equation on $x \in (0, 1)$:

$$-u'' - k^2u = f, \qquad k \in \mathbf{R}$$

with two homogeneous unmixed boundary conditions

$$u(0) = u(1) = 0$$

We identify $p(x) = w(x) = 1$, $q(x) = 0$, $\lambda = k^2$, $a = 0$, $b = 1$, $\alpha = \beta = 0$. In the boundary conditions in (2.38) and (2.39), all coefficients $\alpha_{ij} = 0$, except $\alpha_{11} = \alpha_{23} = 1$. We find that all requirements for SLP1 are satisfied.

■

The operator L in SLP1 has a *formal adjoint*, which we construct by the following procedure. For $u, v \in \mathcal{L}_2(a, b)$, we form

$$\langle Lu, v\rangle = \int_a^b \left\{-\frac{1}{w(x)}\frac{d}{dx}\left[p(x)\frac{du(x)}{dx}\right] + q(x)u(x)\right\} v(x)w(x)dx \tag{2.46}$$

Integrating by parts twice, we obtain

$$\langle Lu, v\rangle = \int_a^b u(x)\left\{-\frac{1}{w(x)}\frac{d}{dx}\left[p(x)\frac{dv(x)}{dx}\right] + q(x)v(x)\right\} w(x)dx$$
$$- \left\{p(x)\left[v(x)\frac{du(x)}{dx} - u(x)\frac{dv(x)}{dx}\right]\right\}\Bigg|_a^b \tag{2.47}$$

We write this result in inner product notation as

$$\langle Lu, v\rangle = \langle u, L^*v\rangle + J(u, v)\Bigg|_a^b \tag{2.48}$$

where $J(u, v)$ is called the *conjunct* and is given by

$$J(u, v) = -p(u'v - uv') \tag{2.49}$$

The operator L^*, produced in the integration by parts operation, is called the formal adjoint to L. We note that

$$L^* = L \tag{2.50}$$

When (2.50) is true, we say that L is *formally self-adjoint.* We conclude, in general, that the Sturm–Liouville operator for SLP1 is formally self-adjoint.

In our search for a solution, or solutions, to (2.35), we shall first assume that the boundary conditions in (2.38) and (2.39) are homogeneous. We then follow with the extension to the inhomogeneous case. Accordingly, if $u(x)$ is to be a solution to (2.35), we require the following restrictions:

a. $u \in \mathcal{L}_2(a, b)$
b. $u \in \mathcal{D}_L$
c. u satisfies two boundary conditions, $B_1(u) = 0$, $B_2(u) = 0$

These restrictions define a linear manifold $\mathcal{M}_L \subset \mathcal{L}_2(a, b)$. The proof is left for the problems. We next consider the function $v(x)$ in (2.48). We place the following restrictions on $v(x)$:

a. $v \in \mathcal{L}_2(a, b)$
b. $v \in \mathcal{D}_{L^*}$
c. v satisfies two *adjoint boundary conditions*, $B_1^*(v) = 0$, $B_2^*(v) = 0$

Since $v(x)$ is unspecified in the original problem statement in (2.35), we are free to choose the adjoint boundary conditions in any manner we wish, consistent with the integration by parts operation in (2.48). We define the adjoint boundary conditions to be those conditions $B_1^*(v) = 0$, $B_2^*(v) = 0$ that, when coupled with the boundary conditions on $u(x)$, result in the vanishing of the conjunct, viz.

$$J(u, v) \Big|_a^b = 0 \tag{2.51}$$

These restrictions on $v(x)$ define a linear manifold $\mathcal{M}_{L^*} \subset \mathcal{L}_2(a, b)$. At present, it is not clear that it is possible to define the adjoint boundary conditions such that (2.51) is satisfied. We next show explicitly the adjoint boundary condition result for the unmixed, initial, and periodic cases.

We have defined the unmixed boundary case in (2.40) and (2.41). For the homogeneous case, they become

$$B_1(u) = \alpha_{11}u(a) + \alpha_{12}u'(a) = 0 \tag{2.52}$$

$$B_2(u) = \alpha_{23}u(b) + \alpha_{24}u'(b) = 0 \tag{2.53}$$

We use these expressions in the conjunct to eliminate $u'(a)$ and $u'(b)$, viz.

$$J(u, v) \Big|_a^b = p(b)u(b)\left[\frac{\alpha_{23}}{\alpha_{24}}v(b) + v'(b)\right] - p(a)u(a)\left[\frac{\alpha_{11}}{\alpha_{12}}v(a) + v'(a)\right] \tag{2.54}$$

In this case, (2.51) is satisfied if we choose the following adjoint boundary conditions:

$$B_1^*(v) = \alpha_{11}v(a) + \alpha_{12}v'(a) = 0 \tag{2.55}$$

$$B_2^*(v) = \alpha_{23}v(b) + \alpha_{24}v'(b) = 0 \tag{2.56}$$

We note that in the unmixed boundary case, the boundary conditions on $v(x)$ in (2.55) and (2.56) are identical to those on $u(x)$ in (2.52) and (2.53). Therefore, for the case of unmixed boundary conditions, the linear manifold $\mathcal{M}_L$ is the same as the linear manifold $\mathcal{M}_{L^*}$. A formally self-adjoint operator with $\mathcal{M}_L = \mathcal{M}_{L^*}$ is said to be *self-adjoint*. We shall find subsequently that self-adjoint problems possess remarkable properties.

For homogeneous initial conditions, we have

$$u(a) = 0 \tag{2.57}$$

$$u'(a) = 0 \tag{2.58}$$

Substitution into the conjunct gives

$$J(u,v)\Big|_a^b = -p(b)\left[u'(b)v(b) - u(b)v'(b)\right] \tag{2.59}$$

In this case, (2.51) is satisfied if we choose adjoint boundary conditions

$$B_1^*(v) = v(b) = 0 \tag{2.60}$$

$$B_2^*(v) = v'(b) = 0 \tag{2.61}$$

We note that for initial conditions, the boundary conditions on $v(x)$ in (2.60) and (2.61) are not the same as those on $u(x)$ in (2.57) and (2.58). Therefore, $\mathcal{M}_L \neq \mathcal{M}_{L^*}$, and the initial condition case is never self-adjoint.

For periodic conditions, we substitute (2.44) and (2.45) into the conjunct to give

$$\begin{aligned} J(u,v)\Big|_a^b &= -p(b)\left[u'(a)v(b) - u(a)v'(b)\right] \\ &\quad + p(a)\left[u'(a)v(a) - u(a)v'(a)\right] \\ &= u'(a)\left[p(a)v(a) - p(b)v(b)\right] \\ &\quad - u(a)\left[p(a)v'(a) - p(b)v'(b)\right] \end{aligned} \tag{2.62}$$

In this case, (2.51) is satisfied if we choose adjoint boundary conditions

$$B_1^*(v) = p(a)v(a) - p(b)v(b) = 0 \tag{2.63}$$

$$B_2^*(v) = p(a)v'(a) - p(b)v'(b) = 0 \tag{2.64}$$

We note that for the general form of L in (2.37) and for periodic conditions, the boundary conditions on $v(x)$ in (2.63) and (2.64) are not the same as those on $u(x)$ in (2.44) and (2.45). However, if the operator L is such that $p(a) = p(b)$, the conditions are identical and the problem becomes self-adjoint.

To produce the solution to SLP1 by the Green's function method, we define two auxiliary problems: the *Green's function problem* and the *adjoint Green's function problem*. The Green's function problem is defined as follows:

$$L_\lambda g(x,\xi) = \frac{\delta(x-\xi)}{w(x)}, \qquad a < \xi < b \tag{2.65}$$

$$B_1(g) = 0 \tag{2.66}$$

$$B_2(g) = 0 \tag{2.67}$$

where $w(x)$ is the weight function defined in (2.25) and (2.27) and appearing in the inner product definition in (2.34). We note that, by definition, the boundary conditions on g are identical to the boundary conditions on u. The adjoint Green's function problem is defined as follows:

$$L_\lambda h(x, \xi) = \frac{\delta(x-\xi)}{w(x)}, \qquad a < \xi < b \tag{2.68}$$

$$B_1^*(h) = 0 \tag{2.69}$$

$$B_2^*(h) = 0 \tag{2.70}$$

We note that, by definition, the boundary conditions on h are identical to the boundary conditions on v. We also note that the boundary conditions associated with the Green's function and the adjoint Green's function are *always* homogeneous.

The solution to SLP1 by the Green's function method is obtained by taking the inner product of $L_\lambda u$ with h, viz.

$$\langle L_\lambda u, h\rangle = \langle u, L_\lambda h\rangle + J(u, h)\Big|_{x=a}^{x=b} \tag{2.71}$$

where the integrations indicated by the inner products are with respect to x. From (2.49), the conjunct $J(u, h)$ is given by

$$J(u, h) = -p(x)\left[\frac{du(x)}{dx}h(x, \xi) - u(x)\frac{dh(x, \xi)}{dx}\right] \tag{2.72}$$

Substitution of (2.35) and (2.68) into (2.71) gives

$$u(\xi) = \langle f, h\rangle - J(u, h)\Big|_a^b \tag{2.73}$$

or, explicitly,

$$u(\xi) = \int_a^b f(x)h(x, \xi)w(x)dx + \left\{p(x)\left[\frac{du(x)}{dx}h(x, \xi) - u(x)\frac{dh(x, \xi)}{dx}\right]\right\}\Bigg|_{x=a}^{x=b} \tag{2.74}$$

Equation (2.74) is the formal solution to SLP1, provided that we can solve the adjoint Green's function problem, given in (2.68)–(2.70). For homogeneous boundary conditions $B_1(u) = 0$, $B_2(u) = 0$, the selection

$B_1^*(h) = 0$, $B_2^*(h) = 0$ reduces the second term in (2.74) to zero. The extension to the inhomogeneous case, however, is now available. We simply apply the given boundary conditions $B_1(u) = \alpha$, $B_2(u) = \beta$ in conjunction with the adjoint boundary conditions $B_1^*(h) = 0$, $B_2^*(h) = 0$. We illustrate these results with an example.

EXAMPLE 2.7 Consider the following differential equation on $x \in (a, b)$:

$$-u'' - \lambda u = f$$

with boundary conditions

$$u(a) = \alpha$$

$$u'(a) = \beta$$

In this case, $p(x) = w(x) = 1$, and (2.74) yields

$$u(\xi) = \int_a^b f(x)h(x,\xi)dx + \alpha\frac{dh(a,\xi)}{dx} - \beta h(a,\xi)$$

where we have applied the boundary conditions $u(a) = \alpha$, $u'(a) = \beta$ and the adjoint boundary conditions

$$h(b,\xi) = \frac{dh(b,\xi)}{dx} = 0$$

Note that for homogeneous boundary conditions, $\alpha = \beta = 0$, the conjunct vanishes and

$$u(\xi) = \int_a^b f(x)h(x,\xi)dx$$

■

We should note that in (2.74) and in Example 2.7, the solution is given in terms of the variable ξ, with x as the dummy variable of integration. This notation causes no difficulty since ξ simply refers to a point of evaluation of u on the interval (a, b). We shall subsequently obtain the solution u in terms of x with ξ as the dummy integration variable by a simple interchange of x and ξ. The reader is cautioned, however, to withhold performing this step until after explicit evaluation of the adjoint Green's function. This evaluation is the next subject for discussion.

We now show that it is never necessary to find the adjoint Green's function $h(x, \xi)$ directly from (2.68)–(2.70). Indeed, if we determine the Green's function $g(x, \xi)$, defined by (2.65)–(2.67), the adjoint Green's function follows immediately. The details follow. We form

$$\langle L_\lambda g(x,\xi), h(x,\xi')\rangle = \langle g(x,\xi), L_\lambda h(x,\xi')\rangle + J(g,h)\Big|_{x=a}^{x=b} \tag{2.75}$$

where integrations in the inner products are with respect to x. Application of (2.66), (2.67), (2.69), and (2.70) reduces the conjunct to zero, viz.

$$J(g,h)\Big|_{x=a}^{x=b} = 0$$

Using this result and (2.65) and (2.68) in (2.75) gives

$$h(\xi,\xi') = g(\xi',\xi)$$

A simple variable change gives

$$h(x,\xi) = g(\xi,x) \tag{2.76}$$

We conclude that, if we can find the Green's function $g(x,\xi)$, the adjoint Green's function $h(x,\xi)$ follows immediately by an interchange of x and ξ. Substitution of $h(x,\xi)$ into (2.74) completes the solution to SLP1.

A further simplification occurs if the Green's function problem is self-adjoint. In this case, the operator and boundary conditions for the Green's function and the adjoint Green's function are identical, and we must have $h = g$. Therefore,

$$h(x,\xi) = g(x,\xi) = g(\xi,x) \qquad \text{(self-adjoint case)} \tag{2.77}$$

When $g(x,\xi) = g(\xi,x)$, the Green's function g is said to be *symmetric*. Substituting (2.77) into (2.74) gives, for the self-adjoint case,

$$\begin{aligned} u(\xi) = &\int_a^b f(x)g(x,\xi)w(x)dx \\ &+ \left\{p(x)\left[\frac{du(x)}{dx}g(x,\xi) - u(x)\frac{dg(x,\xi)}{dx}\right]\right\}\Bigg|_{x=a}^{x=b} \end{aligned} \tag{2.78}$$

We note that in the self-adjoint case, we have produced the useful result that it is unnecessary to consider any aspect of the adjoint problem. Indeed, (2.78) involves the Green's function $g(x,\xi)$ rather than the adjoint Green's function $h(x,\xi)$.

The only remaining step in the solution to SLP1 is the specific determination of $g(x,\xi)$. The differential equation that describes the Green's function is given by (2.65). We write this equation for $x \neq \xi$ as follows:

$$L_\lambda g(x,\xi) = 0, \qquad x \neq \xi \tag{2.79}$$

This homogeneous second-order equation can be solved in the following two regions:

a. Region 1: $a < x < \xi$

b. Region 2: $\xi < x < b$

Since the equation is of second order, the solution in Region 1 will contain two as yet undetermined coefficients. In Region 2, the solution will contain two additional undetermined coefficients. These four coefficients require four constraints on the Green's function for their determination. The conditions $B_1(g) = 0$ and $B_2(g) = 0$ provide two constraints. The remaining two are provided by conditions joining together the two regions at $x = \xi$. Recall that in seeking a solution $u(x)$ to $(L - \lambda)u = f$, we have required that $u(x)$ be twice differentiable. In the solution to (2.65), however, we relax this requirement so that the Green's function is required to be simply differentiable on the interval $a < x < b$. This relaxation is logical since the second differentiation of the Green's function produces a singularity function $\delta(x - \xi)$ at $x = \xi$. Since a differentiable function is continuous, the third constraint on the Green's function is that it must be continuous at $x = \xi$. For the fourth constraint, we write explicitly the Sturm–Liouville operator in (2.65), viz.

$$\left\{-\frac{1}{w(x)}\frac{d}{dx}\left[p(x)\frac{d}{dx}\right] + q(x) - \lambda\right\} g(x, \xi) = \frac{\delta(x - \xi)}{w(x)} \tag{2.80}$$

We multiply by $w(x)$ and integrate over the region $(\xi - \epsilon, \xi + \epsilon)$ to give

$$-\int_{\xi-\epsilon}^{\xi+\epsilon} \frac{d}{dx}\left(p\frac{dg}{dx}\right) dx + \int_{\xi-\epsilon}^{\xi+\epsilon} (q - \lambda)gw dx = 1 \tag{2.81}$$

In the second integral, since q, g, and w are continuous, $(q - \lambda)gw$ is continuous over the interval $[\xi - \epsilon, \xi + \epsilon]$. Since the interval is closed and bounded, the continuous function $(q - \lambda)gw$ is bounded on the interval. Let M be the upper bound on $|(q - \lambda)gw|$. Then,

$$\left|\int_{\xi-\epsilon}^{\xi+\epsilon} (q - \lambda)gw dx\right| \leq 2\epsilon M$$

and therefore,

$$\lim_{\epsilon \to 0}\left|\int_{\xi-\epsilon}^{\xi+\epsilon} (q - \lambda)gw dx\right| = 0$$

Performing the integration in the first integral in (2.81) and taking the limit as $\epsilon \to 0$ gives the fourth constraint, viz.

$$\left.\frac{dg}{dx}\right|_{\xi^+} - \left.\frac{dg}{dx}\right|_{\xi^-} = -\frac{1}{p(\xi)} \tag{2.82}$$

where we have used the continuity of $p(x)$ at $x = \xi$. Typically, our notation ξ^- indicates the limit as $x \to \xi$ from below. We collect the characteristics of the Green's function $g(x, \xi)$ that allow for its determination as follows:

$$L_\lambda g = 0, \qquad x \neq \xi \tag{2.83}$$

$$B_1(g) = 0 \tag{2.84}$$

$$B_2(g) = 0 \tag{2.85}$$

$$g|_{\xi^-} = g|_{\xi^+} \tag{2.86}$$

$$\left.\frac{dg}{dx}\right|_{\xi^+} - \left.\frac{dg}{dx}\right|_{\xi^-} = -\frac{1}{p(\xi)} \tag{2.87}$$

We summarize the procedure for determining the Green's function as follows:

1. Solve (2.83) for $x < \xi$ and for $x > \xi$. The result will contain four as yet undetermined coefficients.
2. Apply the boundary conditions indicated in (2.84) and (2.85). These two conditions will result in determination of two of the four coefficients.
3. Apply the *continuity condition* (2.86) and the *jump condition* (2.87). These two conditions will result in the determination of the final two coefficients.

We demonstrate the procedure in several examples.

EXAMPLE 2.8 Consider the following Green's function problem on $x \in (0, b)$:

$$-\frac{d^2g}{dx^2} - k^2 g = \delta(x - \xi) \tag{2.88}$$

$$g(0, \xi) = \frac{dg(0, \xi)}{dx} = 0 \tag{2.89}$$

We shall solve for the Green's function $g(x, \xi)$ by using (2.83)–(2.87). For $x \neq \xi$, (2.88) becomes

$$-\frac{d^2g}{dx^2} - k^2 g = 0, \qquad x \neq \xi \tag{2.90}$$

A solution to (2.90) can be written for the two regions bisected by $x = \xi$ as follows:

$$g(x, \xi) = \begin{cases} A \sin kx + B \cos kx, & x < \xi \\ C \sin kx + D \cos kx, & x > \xi \end{cases} \tag{2.91}$$

This result can be verified by substituting (2.91) into (2.90) to show that the differential equation is satisfied. We apply the two boundary conditions, given explicitly by (2.89). The result is $A = B = 0$, so that

$$g(x, \xi) = \begin{cases} 0, & x < \xi \\ C \sin kx + D \cos kx, & x > \xi \end{cases} \tag{2.92}$$

The remaining two coefficients C, D are evaluated by applying the continuity and jump conditions. Continuity gives

$$C \sin k\xi + D \cos k\xi = 0$$

Jump gives

$$k(C \cos k\xi - D \sin k\xi) = -1$$

Solving simultaneously gives

$$C = -\frac{\cos k\xi}{k}$$

$$D = \frac{\sin k\xi}{k}$$

Substitution into (2.92) and application of a trigonometric identity yields

$$g(x, \xi) = \begin{cases} 0, & x < \xi \\ \dfrac{\sin k(\xi - x)}{k}, & x > \xi \end{cases} \tag{2.93}$$

It is instructive to verify that (2.93) satisfies the requirements for the Green's function given in (2.83)–(2.87). The details are left for the problems.

■

EXAMPLE 2.9 Consider the following Green's function problem on $x \in (0, a)$:

$$-\frac{d^2 g}{dx^2} - k^2 g = \delta(x - \xi) \tag{2.94}$$

$$g(0, \xi) = g(a, \xi) = 0 \tag{2.95}$$

Proceeding as in Example 2.8, we obtain

$$g(x, \xi) = \begin{cases} A \sin kx + B \cos kx, & x < \xi \\ C \sin k(a - x) + D \cos k(a - x), & x > \xi \end{cases} \tag{2.96}$$

Note that the form of solution for $x > \xi$ is chosen so that the arguments for the sine and cosine are equal to zero at the boundary $x = a$. This form satisfies (2.90), while making the evaluation of the undetermined coefficients C and D easier. (The reader should verify that selection of the form $C \sin kx + D \cos kx$ would yield the same result; but the process of coefficient evaluation would be more complicated.) Applying the boundary conditions given in (2.95), we obtain $B = D = 0$, and

$$g(x, \xi) = \begin{cases} A \sin kx, & x < \xi \\ C \sin k(a - x), & x > \xi \end{cases} \tag{2.97}$$

The remaining two coefficients C, D are evaluated by applying the continuity and jump conditions. The results are

$$C = \frac{\sin k\xi}{k \sin ka} \tag{2.98}$$

$$A = \frac{\sin k(a - \xi)}{k \sin ka} \tag{2.99}$$

Substitution into (2.97) yields

$$g(x, \xi) = \frac{1}{k \sin ka} \begin{cases} \sin k(a - \xi) \sin kx, & x < \xi \\ \sin k(a - x) \sin k\xi, & x > \xi \end{cases} \tag{2.100}$$

Note that the Green's function derived in (2.100) is symmetric, $g(x, \xi) = g(\xi, x)$, a result that we anticipate from the unmixed boundary conditions in (2.95) and the self-adjoint property in (2.77).

■

We next summarize the steps for solving the differential equation $L_\lambda u = f$ by the Green's function method. We distinguish two cases.

Nonself-Adjoint Green's Function Problem

1. Write the solution in the form given by (2.74).
2. Substitute the boundary conditions $B_1(u) = \alpha$, $B_2(u) = \beta$ into (2.74).
3. Substitute the adjoint boundary conditions $B_1^*(h) = 0$, $B_2^*(h) = 0$ into (2.74).
4. Solve the Green's function problem given by (2.65)–(2.67).
5. Obtain the adjoint Green's function through (2.76) and substitute into (2.74).
6. Interchange the variables x and ξ in (2.74).

Self-Adjoint Green's Function Problem

1. Write the solution in the form given by (2.78).
2. Substitute the boundary conditions $B_1(u) = \alpha$, $B_2(u) = \beta$ into (2.78).
3. Substitute the boundary conditions $B_1(g) = 0$, $B_2(g) = 0$ into (2.78).
4. Solve the Green's function problem given by (2.65)–(2.67) and substitute into (2.78).
5. Interchange the variables x and ξ in (2.78).

We remark again that, in the self-adjoint case, there is no necessity for considering any aspect of the adjoint problem. We next illustrate these procedures with some examples.

EXAMPLE 2.10 Consider the following differential equation on $x \in (0, b)$:

$$-u'' - k^2 u = f$$

with boundary conditions

$$u(0) = \alpha$$
$$u'(0) = \beta$$

These boundary conditions define an initial value problem, which we know is never self-adjoint. We therefore use (2.74) and obtain

$$u(\xi) = \int_0^b f(x)h(x,\xi)dx + \alpha\frac{dh(0,\xi)}{dx} - \beta h(0,\xi) \tag{2.101}$$

where the adjoint Green's function equation is given by

$$-\frac{d^2h}{dx^2} - k^2 h = \delta(x - \xi) \tag{2.102}$$

and where we have used the adjoint boundary conditions

$$h(b,\xi) = \frac{dh(b,\xi)}{dx} = 0 \tag{2.103}$$

As shown in (2.76), we can obtain the adjoint Green's function $h(x, \xi)$ directly from the Green's function problem, given in this case by

$$-\frac{d^2g}{dx^2} - k^2 g = \delta(x - \xi) \tag{2.104}$$

$$g(0,\xi) = \frac{dg(0,\xi)}{dx} = 0 \tag{2.105}$$

But, we have obtained this Green's function in (2.93), Example 2.8, as follows:

$$g(x,\xi) = \begin{cases} 0, & x < \xi \\ \dfrac{\sin k(\xi - x)}{k}, & x > \xi \end{cases} \tag{2.106}$$

Application of (2.76) yields the adjoint Green's function, viz.

$$h(x,\xi) = \begin{cases} 0, & x > \xi \\ \dfrac{\sin k(x - \xi)}{k}, & x < \xi \end{cases} \tag{2.107}$$

From (2.107), we also obtain

$$h(0,\xi) = -\frac{\sin k\xi}{k} \tag{2.108}$$

$$\frac{dh(0,\xi)}{dx} = \cos k\xi \tag{2.109}$$

Substitution of (2.107)–(2.109) into (2.101) gives

$$u(\xi) = \int_0^{\xi} f(x)\frac{\sin k(x-\xi)}{k}dx + \beta\frac{\sin k\xi}{k} + \alpha\cos k\xi$$

An interchange of x and ξ yields the final solution, viz.

$$u(x) = \int_0^{x} f(\xi)\frac{\sin k(\xi - x)}{k}d\xi + \beta\frac{\sin kx}{k} + \alpha\cos kx \tag{2.110}$$

It is important to assure that our solution in (2.110) satisfies the differential equation and the boundary conditions. To do so, it is necessary to twice differentiate (2.110). This differentiation requires some care. We note that the integral in (2.110) involves a variable upper limit. To differentiate, we make use of a theorem [10], as follows: "The derivative of the definite integral of a continuous function with respect to the upper limit of integration is equal to the value of the integrand function at this upper limit." In (2.110), however, the variable x occurs not only in the upper limit, but also under the integral sign. To remedy this problem, we write (2.110) as follows:

$$u(x) = \frac{1}{k}\operatorname{Im}\left[e^{-ikx}\int_0^{x} f(\xi)e^{ik\xi}d\xi\right] + \beta\frac{\sin kx}{k} + \alpha\cos kx \tag{2.111}$$

Since real differentiation and the imaginary part operator can be interchanged, it is now straightforward to show that this solution satisfies the original differential equation and the boundary conditions. The details are left for Problem 2.11.

■

EXAMPLE 2.11 Consider the following differential equation on $x \in (0, 1)$:

$$-u'' = f$$

with boundary conditions

$$u(0) = \alpha$$

$$u(1) = 0$$

The associated Green's function problem is

$$-\frac{d^2 g}{dx^2} = \delta(x - \xi)$$

$$g(0, \xi) = g(1, \xi) = 0$$

Since the boundary conditions are unmixed, the Green's function problem is self-adjoint. We therefore use (2.78). After application of the boundary conditions on $u(x)$ and $g(x, \xi)$, we have

$$u(\xi) = \int_0^1 f(x) g(x, \xi) dx + \alpha \frac{dg(0, \xi)}{dx} \tag{2.112}$$

Using the procedure for Green's function evaluation, we find that

$$g(x, \xi) = \begin{cases} (1 - \xi)x, & x < \xi \\ (1 - x)\xi, & x > \xi \end{cases}$$

and

$$\frac{dg(0, \xi)}{dx} = 1 - \xi$$

Substitution into (2.112) followed by an interchange of variables yields

$$u(x) = (1 - x) \int_0^x \xi f(\xi) d\xi + x \int_x^1 (1 - \xi) f(\xi) d\xi + \alpha(1 - x)$$

We leave it to the reader to show that this solution satisfies the differential equation and the boundary conditions.

■

In this section, we have defined the requirements for SLP1 and have given a procedure for its solution by the Green's function method. Note that the parameter λ and the forcing function $f(x)$ were constrained to be real. In many of the interesting problems of electromagnetic theory, λ and $f(x)$ are complex. We consider this case in the next section.

2.5 STURM–LIOUVILLE PROBLEM OF THE SECOND KIND

For the second form of the Sturm–Liouville problem, we consider $(L - \lambda)u = f$ over a finite interval $x \in (a, b)$ and for complex λ and complex f. Since we are now dealing with complex quantities, the Hilbert space $\mathcal{L}_2(a, b)$ is now defined with complex inner product

$$\langle u, v\rangle = \int_a^b u(x)\bar{v}(x)w(x)dx \tag{2.113}$$

for all $u, v \in \mathcal{L}_2(a, b)$. We define the *Sturm–Liouville Problem of the Second Kind*, abbreviated SLP2, as follows:

$$L_\lambda u = f, \qquad -\infty < a < x < b < \infty \tag{2.114}$$

where

$$L_\lambda = L - \lambda \tag{2.115}$$

and where

$$L = -\frac{1}{w(x)}\frac{d}{dx}\left[p(x)\frac{d}{dx}\right] + q(x) \tag{2.116}$$

We impose the following restrictions

a. p, p', q, w are real and continuous for $a \le x \le b$
b. $p(x) > 0$, $w(x) > 0$ for $a \le x \le b$
c. λ is complex and independent of x

In addition, we require $u(x) \in \mathcal{D}_{L_\lambda} \subset \mathcal{L}_2(a, b)$, where $\mathcal{D}_{L_\lambda}$ is the domain of the operator L_λ. Because we are dealing with second-order differential operators, the domain is restricted to those functions that are twice differentiable. Finally, we require that $u(x)$ satisfy two boundary conditions as follows:

$$B_1(u) = \alpha = \alpha_{11}u(a) + \alpha_{12}u'(a) + \alpha_{13}u(b) + \alpha_{14}u'(b) \tag{2.117}$$

$$B_2(u) = \beta = \alpha_{21}u(a) + \alpha_{22}u'(a) + \alpha_{23}u(b) + \alpha_{24}u'(b) \tag{2.118}$$

where the α_{ij} are real. Because of the generalization of λ and $f(x)$ to include complex values, we anticipate that the solution $u(x)$ to (2.114) will be a complex function. Since $u(x)$ is complex, (2.117) and (2.118) will, in general, generate complex values of α and β. We note, however, that the operator L is real; that is,

$$\overline{Lu} = L\bar{u} \tag{2.119}$$

In a similar manner to the SLP1 development, we may show that the operator L in SLP2 is *formally self-adjoint*. Indeed, for $u, v \in \mathcal{L}_2(a, b)$, we find that

$$\begin{aligned}
\langle Lu, v\rangle &= \int_a^b \left\{-\frac{1}{w(x)}\frac{d}{dx}\left[p(x)\frac{du(x)}{dx}\right] + q(x)u(x)\right\} \bar{v}(x)w(x)dx \\
&= \int_a^b u(x)\left\{-\frac{1}{w(x)}\frac{d}{dx}\left[p(x)\frac{d\bar{v}(x)}{dx}\right] + q(x)\bar{v}(x)\right\} w(x)dx \\
&\quad - \left\{p(x)\left[\bar{v}(x)\frac{du(x)}{dx} - u(x)\frac{d\bar{v}(x)}{dx}\right]\right\}\Bigg|_a^b \\
&= \langle u, Lv\rangle + J(u, v)\Bigg|_a^b
\end{aligned} \tag{2.120}$$

where we have used $L\bar{v} = \overline{Lv}$ and where, for SLP2,

$$J(u, v) = -p(u'\bar{v} - u\bar{v}') \tag{2.121}$$

We shall require the same restrictions on u and v as those developed for SLP1. For homogeneous boundary conditions, we place the following restrictions on $u(x)$:

a. $u \in \mathcal{L}_2(a, b)$

b. $u \in \mathcal{D}_L$

c. u satisfies two boundary conditions, $B_1(u) = 0$, $B_2(u) = 0$

We place the following restrictions on $v(x)$:

a. $v \in \mathcal{L}_2(a, b)$

b. $v \in \mathcal{D}_{L^*}$

c. v satisfies two *adjoint boundary conditions*, $B_1^*(v) = 0$, $B_2^*(v) = 0$

We define the adjoint boundary conditions to be those conditions $B_1^*(v) = 0$ $B_2^*(v) = 0$ that, when coupled with the boundary conditions on $u(x)$, result in the vanishing of the conjunct, viz.

$$J(u, v)\Bigg|_a^b = 0 \tag{2.122}$$

We note in (2.121) that the conjunct involves $\bar{v}$, rather than v. However, the conditions on v are easily produced from the conditions on $\bar{v}$. We define the *conjugate adjoint boundary conditions* by

$$\alpha_{11}^*\bar{v}(a) + \alpha_{12}^*\bar{v}'(a) + \alpha_{13}^*\bar{v}(b) + \alpha_{14}^*\bar{v}'(b) = 0 \tag{2.123}$$

$$\alpha_{21}^*\bar{v}(a) + \alpha_{22}^*\bar{v}'(a) + \alpha_{23}^*\bar{v}(b) + \alpha_{24}^*\bar{v}'(b) = 0 \tag{2.124}$$

where the α_{ij}^*'s are chosen so that (2.122) is satisfied. Taking the complex conjugate of (2.123) and (2.124) and noting that the α_{ij}^*'s are real, we obtain

$$\alpha_{11}^*v(a) + \alpha_{12}^*v'(a) + \alpha_{13}^*v(b) + \alpha_{14}^*v'(b) = 0 \tag{2.125}$$

$$\alpha_{21}^*v(a) + \alpha_{22}^*v'(a) + \alpha_{23}^*v(b) + \alpha_{24}^*v'(b) = 0 \tag{2.126}$$

We conclude that, because the α_{ij}'s are real in SLP2, the conjugate adjoint boundary conditions and the adjoint boundary conditions are identical.

We next consider specific conditions that result in L being self-adjoint in SLP2. For the unmixed boundary case, we again have

$$\alpha_{11}u(a) + \alpha_{12}u'(a) = 0 \tag{2.127}$$

$$\alpha_{23}u(b) + \alpha_{24}u'(b) = 0 \tag{2.128}$$

We use these expressions in the conjunct to eliminate $u'(a)$ and $u'(b)$, viz.

$$J(u, v)\Big|_a^b = p(b)u(b)\left[\frac{\alpha_{23}}{\alpha_{24}}\bar{v}(b) + \bar{v}'(b)\right] - p(a)u(a)\left[\frac{\alpha_{11}}{\alpha_{12}}\bar{v}(a) + \bar{v}'(a)\right] \tag{2.129}$$

The conjunct in (2.129) vanishes provided

$$\alpha_{11}\bar{v}(a) + \alpha_{12}\bar{v}'(a) = 0 \tag{2.130}$$

$$\alpha_{23}\bar{v}(b) + \alpha_{24}\bar{v}'(b) = 0 \tag{2.131}$$

which, in SLP2, always implies that

$$\alpha_{11}v(a) + \alpha_{12}v'(a) = 0 \tag{2.132}$$

$$\alpha_{23}v(b) + \alpha_{24}v'(b) = 0 \tag{2.133}$$

We note that, in SLP2, the unmixed boundary case yields boundary conditions on $v(x)$ identical to those on $u(x)$. Therefore, the linear manifold $\mathcal{M}_L$ is the same as the linear manifold $\mathcal{M}_{L^*}$. We conclude that unmixed boundary conditions in SLP2 yield a self-adjoint operator just as in SLP1. We remark that this result depends on the restriction to real α_{ij}. We shall find in the next chapter that this restriction has a dramatic effect on the eigenvalues of the operator L.

EXAMPLE 2.12 We consider characteristics of the following differential equation on $x \in (0, 1)$:

$$(L - \lambda)u = f$$

with $L = -d^2/dx^2$ and with boundary conditions

$$u'(0) - \alpha u(0) = 0$$
$$u(1) = 0$$

where λ and f are complex and α is real. We note that the problem meets all of the requirements for SLP2. In addition, the boundary conditions are unmixed. We therefore conclude that the operator L is self-adjoint. We stress that a different result would have been obtained if $\alpha \in \mathbf{C}$. Indeed, for $u, v \in \mathcal{L}_2(a, b)$, we have

$$\langle Lu, v\rangle = \langle u, Lv\rangle - u'(1)\bar{v}(1) - u(0)\left[\bar{v}'(0) - \alpha\bar{v}(0)\right]$$

where we have applied the boundary conditions on u. The conjugate adjoint boundary conditions that reduce the conjunct to zero are

$$\bar{v}'(0) - \alpha\bar{v}(0) = 0$$
$$\bar{v}(1) = 0$$

Taking the complex conjugate, we have

$$v'(0) - \bar{\alpha}v(0) = 0$$
$$v(1) = 0$$

We conclude that the conditions on v are not the same as the conditions on u, and therefore the operator L is no longer self-adjoint. We shall investigate the distinction between real and complex α in this example again in the next chapter.

■

We next consider homogeneous initial conditions. Following a similar procedure to that in (2.127)–(2.133), we find that the initial condition case is not self-adjoint in SLP2. The details are left for the problems. For periodic conditions in SLP2, a similar procedure shows that the operator L is self-adjoint, provided that $p(a) = p(b)$. Again, the details are left for the problems.

The solution procedure for SLP1 has been given in (2.71)–(2.74). We now show that the solution to SLP2 follows along similar lines, with modification to accommodate complex λ and $f(x)$. We take the inner product of $L_\lambda u$ with the adjoint Green's function h and integrate by parts twice to give

$$\langle L_\lambda u, h\rangle = \langle u, L_\lambda^* h\rangle + J(u,h)\Big|_{x=a}^{x=b} \tag{2.134}$$

where

$$L_\lambda^* = L - \bar{\lambda} \tag{2.135}$$

and

$$J(u,h) = -p(x)\left[\frac{du(x)}{dx}\bar{h}(x,\xi) - u(x)\frac{d\bar{h}(x,\xi)}{dx}\right] \tag{2.136}$$

The adjoint Green's function problem is given by

$$L_\lambda^* h = \frac{\delta(x-\xi)}{w(x)} \tag{2.137}$$

Substitution of (2.137) into (2.134) gives

$$u(\xi) = \langle f, h\rangle - J(u,h)\Big|_{x=a}^{x=b} \tag{2.138}$$

or, explicitly,

$$\begin{aligned} u(\xi) = &\int_a^b f(x)\bar{h}(x,\xi)w(x)dx \\ &+ \left\{p(x)\left[\frac{du(x)}{dx}\bar{h}(x,\xi) - u(x)\frac{d\bar{h}(x,\xi)}{dx}\right]\right\}\Big|_{x=a}^{x=b} \end{aligned} \tag{2.139}$$

We note that (2.139) is the solution to SLP2, provided that we can determine the *conjugate adjoint Green's function* $\bar{h}(x,\xi)$. Taking the complex conjugate of both sides of (2.137), we obtain

$$\overline{L_\lambda^* h(x,\xi)} = L_\lambda \bar{h}(x,\xi) = \frac{\delta(x-\xi)}{w(x)} \tag{2.140}$$

$$B_1^*(\bar{h}) = 0 \tag{2.141}$$

$$B_2^*(\bar{h}) = 0 \tag{2.142}$$

Similar to the SLP1 case, the selection of the boundary conditions proceeds as follows. We first assume that the boundary conditions on u are homogeneous. Then, $B_1^*(\bar{h})$ and $B_2^*(\bar{h})$ are chosen such that the conjunct in (2.138) vanishes. The extension to the inhomogeneous case, however, is now available. We simply apply the given boundary conditions $B_1(u) = \alpha$,

$B_2(u) = \beta$ in conjunction with the conjugate adjoint boundary conditions $B_1^*(\bar{h}) = 0$, $B_2^*(\bar{h}) = 0$.

As in the case of SLP1, we can show that it is never necessary to find the conjugate adjoint Green's function directly. Indeed, we form

$$\langle L_\lambda g(x,\xi), h(x,\xi')\rangle = \langle g(x,\xi), L_\lambda^* h(x,\xi')\rangle + J(g,h)\Big|_{x=a}^{x=b} \tag{2.143}$$

from which

$$\bar{h}(\xi,\xi') = g(\xi',\xi)$$

or, with a variable change,

$$\bar{h}(x,\xi) = g(\xi,x) \tag{2.144}$$

We shall always proceed to find the Green's function $g(x,\xi)$, and then produce the conjugate adjoint Green's function $\bar{h}(x,\xi)$ from (2.144). Substitution of $\bar{h}(x,\xi)$ into (2.139) completes the solution to SLP2.

A further simplification occurs when the operator L is self-adjoint. We have

$$L_\lambda g(x,\xi) = \frac{\delta(x-\xi)}{w(x)} \tag{2.145}$$

$$B_1(g) = 0 \tag{2.146}$$

$$B_2(g) = 0 \tag{2.147}$$

and

$$L_\lambda \bar{h}(x,\xi) = \frac{\delta(x-\xi)}{w(x)} \tag{2.148}$$

$$B_1(\bar{h}) = 0 \tag{2.149}$$

$$B_2(\bar{h}) = 0 \tag{2.150}$$

The fact that the boundary conditions on $\bar{h}$ are identical to the boundary conditions on g is deduced as follows. First, the conditions on g and h are identical from the self-adjoint property of L; second, the conditions on h and $\bar{h}$ are always identical in SLP2. We conclude from (2.145)–(2.150) that $g(x,\xi) = \bar{h}(x,\xi)$, and therefore, using (2.144), we find that

$$g(x,\xi) = \bar{h}(x,\xi) = g(\xi,x) \qquad \text{(self-adjoint case)} \tag{2.151}$$

Substitution into (2.139) gives, for the self-adjoint case,

$$u(\xi) = \int_a^b f(x)g(x,\xi)w(x)dx + \left\{ p(x)\left[\frac{du(x)}{dx}g(x,\xi) - u(x)\frac{dg(x,\xi)}{dx}\right]\right\}\Bigg|_{x=a}^{x=b} \tag{2.152}$$

We shall summarize the steps for solving SLP2 by the Green's function method. We distinguish two cases.

Nonself-Adjoint Green's Function Problem

1. Write the solution in the form given by (2.139).
2. Substitute the boundary conditions $B_1(u) = \alpha$, $B_2(u) = \beta$ into (2.139).
3. Substitute the adjoint boundary conditions $B_1^*(\bar{h}) = 0$, $B_2^*(\bar{h}) = 0$ into (2.139).
4. Solve the Green's function problem given by (2.65)–(2.67).
5. Obtain the conjugate adjoint Green's function $\bar{h}$ through (2.144) and substitute into (2.139).
6. Interchange the variables x and ξ in (2.139).

Self-Adjoint Green's Function Problem

1. Write the solution in the form given by (2.152).
2. Substitute the boundary conditions $B_1(u) = \alpha$, $B_2(u) = \beta$ into (2.152).
3. Substitute the boundary conditions $B_1(g) = 0$, $B_2(g) = 0$ into (2.152).
4. Solve the Green's function problem given by (2.65)–(2.67) and substitute into (2.152).
5. Interchange the variables x and ξ in (2.152).

It is interesting and extremely useful to note that for L self-adjoint, the procedure for obtaining the solution to SLP1 and SLP2 is identical. Indeed, we have proved that in both of these cases, we may obtain the solution in terms of the Green's function $g(x, \xi)$ rather than the adjoint Green's function (SLP1) or the conjugate adjoint Green's function (SLP2). Specifically, (2.78) and (2.151) are identical. It is only in the cases of nonself-adjoint operators where we use the adjoint Green's function (SLP1) or the conjugate adjoint Green's function (SLP2), respectively. We illustrate these ideas in the following examples.

EXAMPLE 2.13 Consider the following differential equation on $x \in (0, a)$:

$$(L - \lambda)u = f$$

$$L = -\frac{d^2}{dx^2}$$

with boundary conditions

$$u'(0) = 0$$

$$u'(a) = 0$$

where f and λ are complex. The problem is of class SLP2. Since the boundary conditions are unmixed, the operator L is self-adjoint with respect to the complex inner product

$$\langle u, v \rangle = \int_0^a u(x)\bar{v}(x)dx$$

Because of the self-adjoint property of L, the form of the solution is given by (2.152). In this case,

$$u(\xi) = \int_0^a f(x)g(x, \xi)dx$$

The self-adjoint property produces a symmetric Green's function, so that

$$u(x) = \int_0^a f(\xi)g(x, \xi)d\xi$$

where we require the solution to the Green's function problem

$$-\frac{d^2 g}{dx^2} - \lambda g = \delta(x - \xi)$$

with boundary conditions

$$\frac{dg(0, \xi)}{dx} = \frac{dg(a, \xi)}{dx} = 0$$

We form

$$g = \begin{cases} A \cos \sqrt{\lambda} x, & x < \xi \\ B \cos \sqrt{\lambda}(a - x), & x > \xi \end{cases}$$

where we have applied the boundary conditions at $x = 0$ and $x = a$ to eliminate two coefficients. Application of the continuity and jump conditions at $x = \xi$ yields

$$A = -\frac{\cos \sqrt{\lambda}(a - \xi)}{\sqrt{\lambda} \sin \sqrt{\lambda} a}$$

$$B = -\frac{\cos \sqrt{\lambda} \xi}{\sqrt{\lambda} \sin \sqrt{\lambda} a}$$

The Green's function therefore is given by

$$g(x,\xi) = -\frac{1}{\sqrt{\lambda}\sin\sqrt{\lambda}a}\begin{cases}\cos\sqrt{\lambda}x\cos\sqrt{\lambda}(a-\xi), & x<\xi\\ \cos\sqrt{\lambda}\xi\cos\sqrt{\lambda}(a-x), & x>\xi\end{cases}$$

As expected from the self-adjoint property, the Green's function is symmetric.

■

EXAMPLE 2.14 Consider the following differential equation on $x \in (0, b)$:

$$-u'' - k^2 u = f$$

with boundary conditions

$$u(0) = \alpha$$

$$u'(0) = \beta$$

where k, α, β, f are complex. The problem is of class SLP2. It is identical to Example 2.10 except for the extension to complex k, α, β, f. Since the boundary conditions define an initial value problem, it is not self-adjoint. We therefore use (2.139) and find that

$$u(\xi) = \int_0^b f(x)\bar{h}(x,\xi)dx + \alpha\frac{d\bar{h}(0,\xi)}{dx} - \beta\bar{h}(0,\xi)$$

where the conjugate adjoint Green's function equation is given by

$$-\frac{d^2\bar{h}}{dx^2} - k^2\bar{h} = \delta(x-\xi)$$

and where we have used the conjugate adjoint boundary conditions

$$\bar{h}(b,\xi) = \frac{d\bar{h}(b,\xi)}{dx} = 0$$

We note that the conjugate adjoint Green's function problem is identical to the adjoint Green's function problem in Example 2.10. The solution therefore proceeds identically, and we produce the following result:

$$u(x) = \int_0^x f(\xi)\frac{\sin k(\xi - x)}{k}d\xi + \beta\frac{\sin kx}{k} + \alpha\cos kx$$

Although the form of solution is the same as in Example 2.10, complex k, α, β, f produces a complex solution $u(x)$.

■

In our study of SLP1 and SLP2 problems, SLP1 could properly be considered as SLP2, with the specialization that all quantities are real. We now take that point of view and classify all problems so far studied in this chapter as SLP2.

Green's function problems classified as SLP2 do not nearly exhaust all of the cases of practical interest. There are many problems of interest in electromagnetics that do not satisfy the requirements of SLP2. Such problems are classified SLP3 and are considered in the next section.

2.6 STURM–LIOUVILLE PROBLEM OF THE THIRD KIND

In defining the *Sturm–Liouville Problem of the Third Kind*, abbreviated SLP3, we again consider the following differential equation:

$$L_\lambda u = f, \qquad a < x < b \tag{2.153}$$

where

$$L_\lambda = L - \lambda, \qquad \lambda \in \mathbf{C} \tag{2.154}$$

and where

$$L = -\frac{1}{w(x)}\frac{d}{dx}\left[p(x)\frac{d}{dx}\right] + q(x) \tag{2.155}$$

In SLP2, we demanded that the interval (a, b) be finite and that the coefficients in (2.155) satisfy the following conditions:

a. p, p', q, w are real and continuous for $a \le x \le b$
b. $p(x) > 0$, $w(x) > 0$ for $a \le x \le b$

If the interval (a, b) is not finite, or if any of the above conditions on the coefficients is violated, the problem is SLP3. In the mathematical literature, SLP2 problems are termed *regular* Sturm–Liouville problems, while SLP3 problems are termed *singular* [11]. We consider the SLP3 problem in Hilbert space $\mathcal{L}_2(a, b)$ with inner product

$$\langle f, g\rangle = \int_a^b f(x)\overline{g}(x)w(x)dx \tag{2.156}$$

There are several classes of SLP3 problems that are important in electromagnetic applications, defined by the following situations:

1. The interval is semi-infinite. In this problem, there is a *singular point* as $x \to \infty$.

2. The interval is finite, but $p(x) = 0$ at an endpoint. In this problem, there is a singular point at the endpoint where $p(x)$ vanishes.
3. The interval is $(-\infty, \infty)$. In this problem, there are singular points as $x \to \pm\infty$.
4. The interval is semi-infinite, and $p(x)$ vanishes at the finite endpoint. In this problem, there are singular points as $x \to \infty$ and at the finite endpoint.

We shall classify singular problems by considering the homogeneous equation associated with (2.153), viz.

$$L_\lambda u = 0 \tag{2.157}$$

According to Weyl's theorem [12]:

1. If for a particular value of λ, every u that is a solution to (2.157) is in $\mathcal{L}_2(a, b)$, then for all λ, every u is in $\mathcal{L}_2(a, b)$.
2. For every λ with $\text{Im}(\lambda) \neq 0$, there exists at least one $u \in \mathcal{L}_2(a, b)$.

We omit the proof of this theorem and refer the reader to [12]. The theorem effectively divides singular problems into two mutually exclusive cases [13]:

1. The *limit circle* case: All solutions u are in $\mathcal{L}_2(a, b)$ for all λ.
2. The *limit point* case: There is either one solution or no solutions in $\mathcal{L}_2(a, b)$, according to the following:
   a. If $\text{Im}(\lambda) \neq 0$, there exists exactly one solution u in $\mathcal{L}_2(a, b)$.
   b. If $\text{Im}(\lambda) = 0$, there is either one solution or no solutions in $\mathcal{L}_2(a, b)$.

We note that we can determine the limit point or limit circle case by examination of the solutions to (2.157) at a single value of λ. If all solutions u are in $\mathcal{L}_2(a, b)$, the limit circle case applies. If not, by exclusion, the limit point case applies.

EXAMPLE 2.15 Consider the following differential equation in $\mathcal{L}_2(0, \infty)$:

$$\left(-\frac{d^2}{dx^2} - \lambda\right) u = 0 \tag{2.158}$$

The endpoint $x = 0$ is a regular point. The endpoint $x \to \infty$ is a singular point. To determine the limit point or limit circle case, we examine solutions to (2.158) for $\lambda = 0$. Two linearly independent solutions are $u_1 = 1$ and $u_2 = x$, neither of which is absolutely square integrable over $(0, \infty)$. Therefore, the solutions are not in $\mathcal{L}_2(0, \infty)$ and we have the limit point case.

■

EXAMPLE 2.16 Consider Bessel's equation of order zero in $\mathcal{L}_2(0, a)$:

$$\left[-\frac{1}{x}\frac{d}{dx}\left(x\frac{d}{dx}\right) - \lambda\right]u = 0, \qquad 0 < a < \infty \tag{2.159}$$

The endpoint $x = a$ is a regular point. Since $p(x) = x = 0$ at $x = 0$, the endpoint $x = 0$ is a singular point. To determine the limit point or limit circle case, we examine two linearly independent solutions to (2.159) for $\lambda = 0$, namely, $u_1 = 1$ and $u_2 = \log x$. Although u_2 is logarithmically singular at $x = 0$, both u_1 and u_2 are absolutely square integrable over $(0, a)$. Therefore, both solutions are in $\mathcal{L}_2(0, a)$, and we have the limit circle case.

■

EXAMPLE 2.17 Consider Bessel's equation of order zero in $\mathcal{L}_2(0, \infty)$:

$$\left[-\frac{1}{x}\frac{d}{dx}\left(x\frac{d}{dx}\right) - \lambda\right]u = 0 \tag{2.160}$$

Both endpoints are singular points. In such cases, we pick an interior point $x = \xi$ and examine limit point or limit circle conditions on two intervals: $\xi < x < \infty$ and $0 < x < \xi$. From Example 2.16, we have the limit circle case on $0 < x < \xi$. For $\xi < x < \infty$, the endpoint $x = \xi$ is regular and the endpoint $x \to \infty$ is singular. Further, neither $u_1 = 1$ nor $u_2 = \log x$ is absolutely square integrable over (ξ, ∞), and therefore neither is in $\mathcal{L}_2(\xi, \infty)$. We conclude that we have the limit point case. We say that Bessel's equation of order zero in $\mathcal{L}_2(0, \infty)$ is in the limit circle case at $x = 0$ and the limit point case as $x \to \infty$.

■

The method of construction of the Green's function for SLP3 problems is directly related to the limit point and limit circle classifications. We shall proceed by considering a few examples, and follow with some conclusions and generalizations.

EXAMPLE 2.18 Consider the following Green's function problem on the interval $x \in (0, \infty)$:

$$-\frac{d^2 g}{dx^2} - \lambda g = \delta(x - \xi), \qquad \lambda \in \mathbf{C} \tag{2.161}$$

with the boundary condition

$$g(0, \xi) = 0$$

In the beginning, we shall not assign a boundary condition as $x \to \infty$. However, the method for dealing with this deficiency will emerge as we proceed. From Example 2.15, we have the limit point case. We begin the construction of the Green's function by considering solutions to the homogeneous equation for $x \neq \xi$, viz.

$$-\frac{d^2 g}{dx^2} - \lambda g = 0, \qquad x \neq \xi \tag{2.162}$$

Possible forms of solution to this equation are $\sin\sqrt{\lambda}x$, $\cos\sqrt{\lambda}x$, $\exp(i\sqrt{\lambda}x)$, and $\exp(-i\sqrt{\lambda}x)$. Since the problem is a limit point problem, we know from Weyl's Theorem that, if $\mathrm{Im}(\lambda) \neq 0$, there is exactly one solution in $\mathcal{L}_2(0,\infty)$. Our task is to find it. (We shall have no need to consider the case where $\mathrm{Im}(\lambda) = 0$ since we can always approach this case by taking a limit as $\mathrm{Im}(\lambda) \to 0$.) Since neither $\sin\sqrt{\lambda}x$ nor $\cos\sqrt{\lambda}x$ is absolutely square integrable over $(0,\infty)$, neither is in $\mathcal{L}_2(0,\infty)$. Consider the two exponential solution forms. We have

$$\int_0^\infty |e^{-i\sqrt{\lambda}x}|^2 dx = \int_0^\infty e^{2(\mathrm{Im}\sqrt{\lambda})x} dx \tag{2.163}$$

and

$$\int_0^\infty |e^{i\sqrt{\lambda}x}|^2 dx = \int_0^\infty e^{-2(\mathrm{Im}\sqrt{\lambda})x} dx \tag{2.164}$$

Which of these two exponential solution forms is in $\mathcal{L}_2(0,\infty)$ depends on whether $\mathrm{Im}(\sqrt{\lambda})$ is negative or positive. Since λ is a parameter specified in the problem statement, we shall choose for definiteness

$$\mathrm{Im}(\sqrt{\lambda}) < 0 \tag{2.165}$$

With this choice

$$\int_0^\infty |e^{-i\sqrt{\lambda}x}|^2 dx < \infty \tag{2.166}$$

and we conclude that $\exp(-i\sqrt{\lambda}x)$ is the one solution to (2.162) in $\mathcal{L}_2(0,\infty)$. We now proceed with the construction of the Green's function in the usual manner. We write

$$g(x,\xi) = \begin{cases} A\sin\sqrt{\lambda}x + C\cos\sqrt{\lambda}x, & x < \xi \\ Be^{-i\sqrt{\lambda}x} + De^{i\sqrt{\lambda}x}, & x > \xi \end{cases} \tag{2.167}$$

Application of the boundary condition at $x = 0$ results in $C = 0$, with the result

$$g(x,\xi) = \begin{cases} A\sin\sqrt{\lambda}x, & x < \xi \\ Be^{-i\sqrt{\lambda}x} + De^{i\sqrt{\lambda}x}, & x > \xi \end{cases} \tag{2.168}$$

In SLP1 or SLP2 problems, we would next apply a second boundary condition to eliminate another coefficient. In the limit point case in SLP3, however, we replace the second boundary condition with the requirement that the solution be in $\mathcal{L}_2(0, \infty)$. Since $\sin\sqrt{\lambda}x$, as used in (2.168), has support only on $(0, \xi)$, the only part of the solution in (2.168) that is not in $\mathcal{L}_2(0, \infty)$ is $\exp(i\sqrt{\lambda}x)$. We therefore choose $D = 0$ and obtain

$$g(x, \xi) = \begin{cases} A \sin\sqrt{\lambda}x, & x < \xi \\ Be^{-i\sqrt{\lambda}x}, & x > \xi \end{cases} \tag{2.169}$$

We next apply the continuity and jump conditions at $x = \xi$ in the usual manner and obtain

$$A = \frac{e^{-i\sqrt{\lambda}\xi}}{\sqrt{\lambda}} \tag{2.170}$$

$$B = \frac{\sin\sqrt{\lambda}\xi}{\sqrt{\lambda}} \tag{2.171}$$

Substitution of these constants into (2.170) gives

$$g(x, \xi) = \frac{1}{\sqrt{\lambda}} \begin{cases} e^{-i\sqrt{\lambda}\xi} \sin\sqrt{\lambda}x, & x < \xi \\ e^{-i\sqrt{\lambda}x} \sin\sqrt{\lambda}\xi, & x > \xi \end{cases} \tag{2.172}$$

where $\sqrt{\lambda}$ is constrained by (2.165). We note that the Green's function derived in (2.172) is symmetric, $g(x, \xi) = g(\xi, x)$.

■

Example 2.18 suggests the following procedure for dealing with Green's functions associated with problems in the limit point case at one boundary, say $x = b$, and regular at the other boundary. First, we write the solution to the Green's function problem in the usual manner, in terms of four undetermined coefficients, as in (2.167). To determine one of the four coefficients, we apply the boundary condition at the regular endpoint. Next, to determine a second coefficient, we apply the requirement that the solution on the interval $\xi < x < b$ must be in $\mathcal{L}_2(a, b)$. The remaining two coefficients are determined in the usual manner by the continuity and jump conditions at $x = \xi$.

Mathematically, for the limit point case, we may show that for $\text{Im}(\lambda) \neq 0$, the single solution to $L_\lambda u = 0$ in $\mathcal{L}_2(a, b)$ is always obtained simply by invoking the $\mathcal{L}_2$ requirement. In addition, if an unmixed boundary condition is applied at the regular endpoint, this condition, together with the $\mathcal{L}_2$ requirement, renders the problem self-adjoint. *No boundary condition is required at the limit point boundary.* The proof of this crucial result is contained in a review paper by Hajmirzaahmad and Krall [14]. As we have shown, the self-adjoint property results in a symmetric Green's function.

There is an alternate method leading to the determination of the Green's function in Example 2.18 [15]. Indeed, if we invoke in (2.168) the physically reasonable condition that the Green's function vanishes as $x \to \infty$, we produce the same result as we do by invoking the $\mathcal{L}_2(0, \infty)$ requirement. That is, we can invoke a *limit condition*

$$\lim_{x \to \infty} g(x, \xi) = 0$$

in place of the second "boundary" condition in the statement of the problem. This is an appealing procedure since such an unmixed limit condition can be viewed as an extension of the regular boundary condition

$$g(b, \xi) = 0$$

Indeed, consider the Green's function problem

$$-\frac{d^2 g}{dx^2} - \lambda g = \delta(x - \xi)$$

with boundary conditions

$$g(0, \xi) = g(b, \xi) = 0$$

The result in (2.172) can be obtained by solving this problem and then taking the limit as $b \to \infty$. The details are left for the problems.

In summary, for the case of a regular unmixed boundary condition at $x = a$ and the limit point case at $x = b$, the $\mathcal{L}_2$ requirement takes the place of a boundary condition at $x = b$. Furthermore, the problem is self-adjoint. In the case where $b \to \infty$, we may use the alternate procedure of applying a limit condition in place of the $\mathcal{L}_2$ requirement. We remark that it is sufficient to have a procedure that picks out the one $\mathcal{L}_2$ solution required in the mathematical proofs, such as those in [14]. Invoking the limit condition is such a procedure. We consider these ideas further in the following example.

EXAMPLE 2.19 Consider the following differential equation on $x \in (0, \infty)$:

$$-u'' - \lambda u = f \tag{2.173}$$

with boundary condition

$$u(0) = 0$$

where u, f, λ are complex. We choose the inner product

$$\langle u, v\rangle = \int_0^\infty u(x)\bar{v}(x)dx$$

From Example 2.18, we know that this problem is singular in the limit point case as $x \to \infty$. We therefore invoke the limit condition

$$\lim_{x\to\infty} u(x) = 0$$

The problem is self-adjoint and the Green's function is symmetric. We therefore use (2.152) which, specialized to this case, yields

$$u(\xi) = \int_0^\infty f(x)g(x,\xi)dx + \left[\frac{du(x)}{dx}g(x,\xi) - u(x)\frac{dg(x,\xi)}{dx}\right]\Bigg|_{x=0}^{x=\infty} \tag{2.174}$$

We apply the boundary condition and limit condition on $u(x)$ and choose

$$g(0,\xi) = 0$$

$$\lim_{x\to\infty} g(x,\xi) = 0$$

and find that

$$u(\xi) = \int_0^\infty f(x)g(x,\xi)dx$$

Finally, after interchanging x and ξ, we obtain

$$u(x) = \int_0^\infty f(\xi)g(x,\xi)d\xi$$

where the Green's function $g(x,\xi)$ is given by (2.172).

■

We note that the result in (2.174) is an extension to the result for self-adjoint operators in SLP2. Specifically, the arguments for the SLP2 unmixed boundary case given in (2.122)–(2.133) carry over to the SLP3 limit point case at infinity, provided again that the α_{ij}'s are constrained to be real. We may establish this result simply by observing that the arguments in (2.122)–(2.133) are not altered by taking the limit as $a \to -\infty$ or $b \to \infty$, or both. Since the problem is self-adjoint, the Green's function is symmetric, and the result in (2.174) is assured before solving for the specific Green's function. We shall illustrate this important point in an additional example.

EXAMPLE 2.20 Consider the following differential equation on $x \in (-\infty, \infty)$:

$$(L - \lambda)u = f, \qquad \text{Im}(\sqrt{\lambda}) < 0$$

where

$$L = -\frac{d^2}{dx^2}$$

This problem is in the limit point case as $x \to \infty$ and as $x \to -\infty$. Our procedure in dealing with limit points at both ends of the interval along the real line is to pick an interior point $x = \xi$. Since $\text{Im}(\lambda) \neq 0$, there is exactly one solution to $L_\lambda u = 0$ in $\mathcal{L}_2(-\infty, \xi)$ and exactly one solution in $\mathcal{L}_2(\xi, \infty)$. These two solutions to the homogeneous equation form the building blocks for the construction of the Green's function. Hajmirzaahmad and Krall [14] prove the following: For the Sturm–Liouville operator L with the limit point case at both ends of the interval,

1. No boundary conditions need be invoked.
2. L is self-adjoint.

Again, in lieu of the $\mathcal{L}_2$ requirement, we shall invoke limiting conditions, one at each end of the interval, viz.

$$\lim_{x \to -\infty} u(x) = 0$$

$$\lim_{x \to \infty} u(x) = 0$$

We assume that u, λ, f are complex. Since L is self-adjoint, the Green's function is symmetric. The solution to the differential equation is therefore given by

$$u(x) = \int_{-\infty}^{\infty} f(\xi) g(x, \xi) d\xi$$

where the Green's function must satisfy

$$(L - \lambda) g(x, \xi) = \delta(x - \xi)$$

$$\lim_{x \to -\infty} g(x, \xi) = \lim_{x \to \infty} g(x, \xi) = 0$$

We write the solution for the Green's function as

$$g(x, \xi) = \begin{cases} Ae^{-i\sqrt{\lambda}x}, & x > \xi \\ Be^{i\sqrt{\lambda}x}, & x < \xi \end{cases}$$

where

$$\text{Im}(\sqrt{\lambda}) < 0$$

and where we have invoked the two limit conditions. We note that, consistent with the limit point case and $\mathrm{Im}(\lambda) \neq 0$, our two limit conditions have produced exactly one solution in $\mathcal{L}_2(-\infty, \xi)$ and exactly one solution in $\mathcal{L}_2(\xi, \infty)$. Applying the continuity and jump conditions at $x = \xi$, we obtain

$$A = \frac{e^{i\sqrt{\lambda}\xi}}{2i\sqrt{\lambda}}$$

$$B = \frac{e^{-i\sqrt{\lambda}\xi}}{2i\sqrt{\lambda}}$$

Therefore,

$$g(x, \xi) = \frac{1}{2i\sqrt{\lambda}} \begin{cases} e^{-i\sqrt{\lambda}(x-\xi)}, & x > \xi \\ e^{-i\sqrt{\lambda}(\xi-x)}, & x < \xi \end{cases}$$

or, more compactly,

$$g(x, \xi) = \frac{e^{-i\sqrt{\lambda}|x-\xi|}}{2i\sqrt{\lambda}} \tag{2.175}$$

■

We have established in the above paragraphs a procedure for deriving the Green's function in limit point problems. We now turn to a consideration of the limit circle case. We begin with two examples.

EXAMPLE 2.21 Consider the following Green's function problem on $x \in (0, \infty)$:

$$-\frac{1}{x}\left[\frac{d}{dx}\left(x\frac{dg}{dx}\right)\right] - \lambda g = \frac{\delta(x-\xi)}{x}, \qquad \lambda \in \mathbf{C} \tag{2.176}$$

From the results in Example 2.17, we have the limit circle case at $x = 0$ and the limit point case as $x \to \infty$. We begin our construction of the Green's function in the usual manner by considering the homogeneous equation

$$-\frac{1}{x}\left[\frac{d}{dx}\left(x\frac{dg}{dx}\right)\right] - \lambda g = 0, \qquad x \neq \xi$$

which is Bessel's equation of order zero. Solutions can be constructed from linear combinations of the Bessel function $J_0(\sqrt{\lambda}x)$, the Neumann function $Y_0(\sqrt{\lambda}x)$, and the two Hankel functions $H_0^{(1)}(\sqrt{\lambda}x)$ and $H_0^{(2)}(\sqrt{\lambda}x)$. We write

$$g = \begin{cases} AJ_0(\sqrt{\lambda}x) + CY_0(\sqrt{\lambda}x), & x < \xi \\ BH_0^{(2)}(\sqrt{\lambda}x) + DH_0^{(1)}(\sqrt{\lambda}x), & x > \xi \end{cases} \tag{2.177}$$

We may evaluate one of the coefficients in (2.177) by following our procedure for dealing with limit points. We therefore invoke the following limiting condition as $x \to \infty$:

$$\lim_{x\to\infty} g(x,\xi) = 0 \tag{2.178}$$

The asymptotic forms of the two Hankel functions are given by [16]

$$H_0^{(1)}(t) \sim \sqrt{\frac{2}{i\pi t}}e^{it}$$

$$H_0^{(2)}(t) \sim \sqrt{\frac{2i}{\pi t}}e^{-it}$$

These asymptotic forms show that if we constrain $\operatorname{Im}\sqrt{\lambda} < 0$, $H_0^{(1)}(\sqrt{\lambda}x)$ diverges as $x \to \infty$. We therefore set $D = 0$ and obtain

$$g = \begin{cases} AJ_0(\sqrt{\lambda}x) + CY_0(\sqrt{\lambda}x), & x < \xi \\ BH_0^{(2)}(\sqrt{\lambda}x), & x > \xi \end{cases} \tag{2.179}$$

Determining the remaining three coefficients requires three conditions. The continuity and jump conditions at $x = \xi$ will provide two conditions. To produce the third, we consider the limit circle case at $x = 0$. The leading terms in the expansion of the Bessel and Neumann functions are given by [16]

$$J_0(t) = 1 + \cdots$$

$$Y_0(t) = \frac{2}{\pi}\ln\frac{\gamma t}{2} - \cdots$$

where $\ln\gamma$ is Euler's constant. Since we have the limit circle case, we know *a priori* that both of these functions are square integrable over $(0, \xi)$. Therefore, invoking the requirement that the solution be in $\mathcal{L}_2(0, \xi)$ does not evaluate a coefficient, as was the case for limit points. We do have, however, a condition that we can invoke from physical principles. Bessel's equation with forcing function at $x = \xi$ normally results from considerations of the radial dependence in problems in cylindrical coordinates. In such problems, we shall find in Chapter 4 that, based on physical grounds, the solution must remain finite as $x \to 0$. The Neumann function does not meet such a condition at $x = 0$, and we therefore set $C = 0$ and obtain

$$g = \begin{cases} AJ_0(\sqrt{\lambda}x), & x < \xi \\ BH_0^{(2)}(\sqrt{\lambda}x), & x > \xi \end{cases} \tag{2.180}$$

The continuity condition at $x = \xi$ yields

$$AJ_0(\sqrt{\lambda}\xi) = BH_0^{(2)}(\sqrt{\lambda}\xi) \tag{2.181}$$

The jump condition gives

$$\left[B \frac{dH_0^{(2)}(\sqrt{\lambda}x)}{dx} - A \frac{dJ_0(\sqrt{\lambda}x)}{dx} \right]_{x=\xi} = -\frac{1}{\xi} \tag{2.182}$$

Performing the indicated derivatives and solving (2.181) and (2.182) simultaneously for A gives

$$A \left[J_0(\sqrt{\lambda}\xi) H_1^{(2)}(\sqrt{\lambda}\xi) - J_1(\sqrt{\lambda}\xi) H_0^{(2)}(\sqrt{\lambda}\xi) \right] = \frac{H_0^{(2)}(\sqrt{\lambda}\xi)}{\sqrt{\lambda}\xi} \tag{2.183}$$

By a well-known Wronskian relationship [17], we have

$$J_1(t) H_0^{(2)}(t) - J_0(t) H_1^{(2)}(t) = \frac{2}{i\pi t}$$

Using this relation in (2.183), we obtain

$$A = \frac{\pi}{2i} H_0^{(2)}(\sqrt{\lambda}\xi)$$

Substitution of this result into (2.181) gives

$$B = \frac{\pi}{2i} J_0(\sqrt{\lambda}\xi)$$

Therefore,

$$g(x, \xi) = \frac{\pi}{2i} \begin{cases} H_0^{(2)}(\sqrt{\lambda}\xi) J_0(\sqrt{\lambda}x), & x < \xi \\ H_0^{(2)}(\sqrt{\lambda}x) J_0(\sqrt{\lambda}\xi), & x > \xi \end{cases} \tag{2.184}$$

We note that the Green's function is symmetric, $g(x, \xi) = g(\xi, x)$. A useful specialization of the result in (2.184) can be obtained by taking the limit as $\xi \to 0$, with the result

$$g(x, 0) = \frac{\pi}{2i} H_0^{(2)}(\sqrt{\lambda}x) \tag{2.185}$$

■

We have noted in the above example that the Green's function is symmetric. We are led to inquire if the operator that produced the Green's function is self-adjoint. Consider first the case where we have the Sturm–Liouville operator L with the limit circle case at $x = a$ and a regular unmixed boundary condition at $x = b$. This case has been clarified by Kaper, Kwong, and Zettl [18], who have proved the following: The operator L is self-adjoint if $u \in \mathcal{D}_L$, and if:

1. u satisfies an unmixed condition at the regular boundary $x = b$.
2. u exists and is finite as $u \to a$, the limit circle boundary.

In addition, they show that the finiteness condition is mathematically equivalent to the condition

$$\lim_{x \to a} [p(x)u'(x)] = 0$$

(This equivalence is important in making the connection between the physically appealing finiteness condition and the classical Weyl theory.) This important result has been extended [14] to show that the self-adjoint property is retained when the regular point at $x = b$ is replaced by a limit point, as in the previous example. We consider another example.

EXAMPLE 2.22 Consider the following Green's function problem on $x \in (0, \infty)$:

$$-\frac{1}{x^2}\left[\frac{d}{dx}\left(x^2\frac{dg}{dx}\right)\right] - k^2 g = \frac{\delta(x-\xi)}{x^2}, \qquad k \in \mathbf{C} \tag{2.186}$$

This problem is in the limit circle case at $x = 0$ and the limit point case as $x \to \infty$. We therefore invoke a finiteness condition at $x = 0$ and the following limit condition as $x \to \infty$:

$$\lim_{x \to \infty} g(x, \xi) = 0$$

Equation (2.186) is the *spherical Bessel equation* of order zero [19]. Its solution is given by linear combinations of spherical Bessel, spherical Neumann, and spherical Hankel functions. We therefore write

$$g = \begin{cases} Aj_0(kx) + Ch_0^{(2)}(kx), & x < \xi \\ Bh_0^{(2)}(kx) + Dh_0^{(1)}(kx), & x > \xi \end{cases} \tag{2.187}$$

where

$$j_0(kx) = \frac{\sin kx}{kx} \tag{2.188}$$

$$h_0^{(1)}(kx) = \frac{e^{ikx}}{ikx} \tag{2.189}$$

$$h_0^{(2)}(kx) = -\frac{e^{-ikx}}{ikx} \tag{2.190}$$

To preserve finiteness as $x \to 0$, we set $C = 0$. To satisfy the limit condition at infinity, we adopt the constraint

$$\text{Im}(k) < 0 \tag{2.191}$$

and therefore set $D = 0$. With these conditions, we find that

$$g = \begin{cases} A j_0(kx), & x < \xi \\ B h_0^{(2)}(kx), & x > \xi \end{cases} \tag{2.192}$$

The continuity condition at $x = \xi$ yields

$$A j_0(k\xi) = B h_0^{(2)}(k\xi) \tag{2.193}$$

The jump condition gives

$$\left[B \frac{d h_0^{(2)}(kx)}{dx} - A \frac{d j_0(kx)}{dx} \right]_{x=\xi} = -\frac{1}{\xi^2} \tag{2.194}$$

Solving for A in (2.193) and substituting into (2.194), we have

$$B \left[h_0^{(2)}(k\xi) \frac{d j_0(k\xi)}{dx} - \frac{d h_0^{(2)}(k\xi)}{dx} j_0(k\xi) \right] = \frac{j_0(k\xi)}{\xi^2} \tag{2.195}$$

The expression in square brackets in (2.195) is one of many Wronskian expressions involving the spherical Bessel functions. In this case, we find [20] that

$$h_0^{(2)} j_0' - h_0^{(2)'} j_0 = \frac{i}{z^2} \tag{2.196}$$

where all arguments are with respect to z and differentiation is with respect to argument. Using (2.196) in (2.195) gives

$$B = -ik j_0(k\xi) \tag{2.197}$$

Substitution into (2.193) yields

$$A = -ik h_0^{(2)}(k\xi) \tag{2.198}$$

Substitution of (2.197) and (2.198) into (2.192) yields the Green's function

$$g = -ik \begin{cases} j_0(kx) h_0^{(2)}(k\xi), & x < \xi \\ j_0(k\xi) h_0^{(2)}(kx), & x > \xi \end{cases} \tag{2.199}$$

An alternate form can be obtained by using (2.188) and (2.190). We have

$$g = \frac{1}{kx\xi} \begin{cases} e^{-ik\xi} \sin kx, & x < \xi \\ e^{-ikx} \sin k\xi, & x > \xi \end{cases} \tag{2.200}$$

We note that in the limit as $\xi \to 0$, we produce the result

$$\lim_{\xi \to 0} g(x, \xi) = \frac{e^{-ikx}}{x} \tag{2.201}$$

We shall use this important result in Chapter 4.

■

In dealing with the singular point in limit circle cases, we have not been able to evaluate a coefficient by invoking the requirement that the solution be in $\mathcal{L}_2$. Instead, we have invoked a condition based on physical grounds. The self-adjoint property of L results in a Green's function that is symmetric.

The above examples all concern operators that are self-adjoint. We next present an example where the operator manifold contains two unmixed conditions, but the operator is not self-adjoint.

EXAMPLE 2.23 Consider the following differential equation on $x \in (0, \infty)$:

$$-u'' - \lambda u = f \tag{2.202}$$

where

$$u'(0) = \alpha u(0), \qquad \alpha \in \mathbf{C} \tag{2.203}$$

and where we assume that u, λ, f are complex. The boundary $x = 0$ is a regular point. As $x \to \infty$, we have the limit point case. We therefore apply the limit condition

$$\lim_{x \to \infty} u(x) = 0 \tag{2.204}$$

In addition, since α is complex, the problem is not an extension to a finite interval self-adjoint problem. (See Example 2.12 for a discussion.) We therefore proceed using (2.139). Applying the boundary and limit conditions given in (2.203) and (2.204), we obtain

$$u(\xi) = \int_0^\infty f(x)\bar{h}(x, \xi)dx + u(0)\left[\alpha\bar{h}(0, \xi) - \frac{d\bar{h}(0, \xi)}{dx}\right] \tag{2.205}$$

where we have chosen

$$\lim_{x \to \infty} \bar{h}(x, \xi) = 0 \tag{2.206}$$

If we now choose

$$\frac{d\bar{h}(0, \xi)}{dx} = \alpha\bar{h}(0, \xi) \tag{2.207}$$

we obtain

$$u(\xi) = \int_0^\infty f(x)\bar{h}(x,\xi)dx \tag{2.208}$$

We note that the boundary conditions on $\bar{h}(x,\xi)$ are identical to the boundary conditions on $u(x)$. If we recall that the boundary conditions on the Green's function $g(x,\xi)$ are always identical to the boundary conditions on $u(x)$, we find that

$$\bar{h}(x,\xi) = g(x,\xi) \tag{2.209}$$

We therefore have

$$u(\xi) = \int_0^\infty f(x)g(x,\xi)dx \tag{2.210}$$

where we must solve

$$-\frac{dg^2}{dx^2} - \lambda g = \delta(x-\xi) \tag{2.211}$$

$$\frac{dg(0,\xi)}{dx} = \alpha g(0,\xi) \tag{2.212}$$

$$\lim_{x\to\infty} g(x,\xi) = 0 \tag{2.213}$$

We write

$$g = \begin{cases} A\cos\sqrt{\lambda}x + B\sin\sqrt{\lambda}x, & x < \xi \\ Ce^{-i\sqrt{\lambda}x}, & x > \xi \end{cases} \tag{2.214}$$

where we have applied the limit condition as $x \to \infty$ and have chosen

$$\mathrm{Im}(\sqrt{\lambda}) < 0$$

Applying the boundary condition at $x = 0$, we obtain

$$g = \begin{cases} A\left(\cos\sqrt{\lambda}x + \dfrac{\alpha}{\sqrt{\lambda}}\sin\sqrt{\lambda}x\right), & x < \xi \\ Ce^{-i\sqrt{\lambda}x}, & x > \xi \end{cases} \tag{2.215}$$

Invoking the continuity and jump conditions at $x = \xi$ results in

$$A = \frac{e^{-i\sqrt{\lambda}\xi}}{i\sqrt{\lambda}+\alpha}$$

$$C = \frac{\cos\sqrt{\lambda}\xi + \frac{\alpha}{\sqrt{\lambda}}\sin\sqrt{\lambda}\xi}{i\sqrt{\lambda}+\alpha}$$

Therefore, the Green's function is given by

$$g(x,\xi) = \frac{1}{i\sqrt{\lambda}+\alpha}\begin{cases} e^{-i\sqrt{\lambda}\xi}\left(\cos\sqrt{\lambda}x + \dfrac{\alpha}{\sqrt{\lambda}}\sin\sqrt{\lambda}x\right), & x < \xi \\ e^{-i\sqrt{\lambda}x}\left(\cos\sqrt{\lambda}\xi + \dfrac{\alpha}{\sqrt{\lambda}}\sin\sqrt{\lambda}\xi\right), & x > \xi \end{cases} \tag{2.216}$$

We note that, although the operator in this problem is not self-adjoint, we still have

$$g(x,\xi) = g(\xi,x)$$

and, therefore, interchanging x and ξ in (2.210), we have

$$u(x) = \int_0^\infty f(\xi)g(x,\xi)d\xi$$

We remark that if an operator is self-adjoint, the Green's function is symmetric. However, if the operator is not self-adjoint, it does not necessarily follow that the Green's function is not symmetric. This seemingly small distinction has a marked effect on characteristics of eigenvalues, as will be discussed in the next chapter.

■

In certain cases, determination of the limit point or limit circle is dependent on parameters in the differential equation. We illustrate this fact with the following example.

EXAMPLE 2.24 Consider the following Green's function problem associated with Bessel's equation of order ν in $\mathcal{L}_2(0,\infty)$:

$$(L-\lambda)g = \frac{\delta(x-\xi)}{x} \tag{2.217}$$

where

$$L = -\frac{1}{x}\left[\frac{d}{dx}\left(x\frac{d}{dx}\right)\right] + \frac{\nu^2}{x^2} \tag{2.218}$$

and where ν and λ are complex parameters. We assume that

$$\mathrm{Re}(\nu) > 0 \tag{2.219}$$

We define the inner product for the space to be

$$\langle u, v\rangle = \int_0^\infty u\bar{v}x dx \tag{2.220}$$

We note that both endpoints are singular points. Proceeding as in Example 2.17, we pick an interior point $x = \xi$ and examine limit point or limit circle conditions on two intervals $\xi < x < \infty$ and $0 < x < \xi$. For $\lambda = 0$, the homogeneous equation $(L - \lambda)u = 0$ has the two independent solutions

$$u_1 = x^{\nu} \tag{2.221}$$

and

$$u_2 = x^{-\nu} \tag{2.222}$$

Consider the interval $\xi < x < \infty$. We first examine whether u_1 is in $\mathcal{L}_2(\xi, \infty)$. We have

$$\int_{\xi}^{\infty} x^{\nu} x^{\overline{\nu}} x dx = \int_{\xi}^{\infty} x^{2\mathrm{Re}(\nu)+1} dx$$

By (2.219), this integral diverges and u_1 is not in $\mathcal{L}_2(\xi, \infty)$. We therefore have the limit point case as $x \to \infty$ and assign the limit condition

$$\lim_{x\to\infty} g(x, \xi) = 0 \tag{2.223}$$

The situation on the interval $0 < x < \xi$ is more delicate. We first examine whether u_1 is in $\mathcal{L}_2(0, \xi)$. We have

$$\int_{0}^{\xi} x^{\nu} x^{\overline{\nu}} x dx = \int_{0}^{\xi} x^{2\mathrm{Re}(\nu)+1} dx$$

This integral exists when $2\mathrm{Re}(\nu) + 1 > -1$ or $\mathrm{Re}(\nu) > -1$. Replacing ν by $-\nu$, we find that u_2 is in $\mathcal{L}_2(0, \xi)$, provided that $\mathrm{Re}(-\nu) > -1$ or $\mathrm{Re}(\nu) < 1$. Combining these two results, we find that both solutions are in $\mathcal{L}_2(0, \xi)$, provided that $-1 < \mathrm{Re}(\nu) < 1$. We therefore have the limit circle case as $x \to 0$ for $-1 < \mathrm{Re}(\nu) < 1$ and the limit point case otherwise. In either case, we shall demand satisfaction of the physically motivated limit condition

$$\lim_{x\to 0} g(x, \xi) \text{ finite} \tag{2.224}$$

The Green's function problem defined by (2.217) and (2.218) with limiting conditions given by (2.223) and (2.224) yields the solution

$$g(x, \xi) = \frac{\pi}{2i} \begin{cases} H_{\nu}^{(2)}(\sqrt{\lambda}\xi) J_{\nu}(\sqrt{\lambda}x), & x < \xi \\ H_{\nu}^{(2)}(\sqrt{\lambda}x) J_{\nu}(\sqrt{\lambda}\xi), & x > \xi \end{cases} \tag{2.225}$$

where

$$\mathrm{Im}(\sqrt{\lambda}) < 0$$

That this is the solution for $g(x, \xi)$ can be determined by construction in the usual manner. We defer the details until Example 3.6 in the next chapter.

■

The limit point and limit circle cases in SLP3 problems are a subject of continuing interest to mathematicians. We have only considered the portion of the theory of interest to us in our application to electromagnetic boundary value problems. For an in-depth discussion of limit point and limit circle cases, as well as a well-compiled bibliography, the reader is referred to [14] and [21].

We have now completed our discussion of the solution to linear, ordinary, second-order differential equations by the Green's function method. In the next chapter, we shall discuss an alternate method where we shall determine the solution to $(L-\lambda)u = f$ by finding the spectral representation associated with the differential operator L.

PROBLEMS

2.1. For the pulse function $p_\epsilon(x - x_0)$, defined in (2.2), show that

$$p_\epsilon(x - x_0) = p_\epsilon(x_0 - x)$$

2.2. By substituting (2.23)–(2.25) into (2.22), verify that the Sturm–Liouville differential equation is transformed into the general form in (2.21).

2.3. The Chebyshev differential equation is defined on the interval $x \in (-1, 1)$, as follows:

$$-(1 - x^2)u'' + xu' - n^2u = f$$

Transform to Sturm–Liouville form.

2.4. Transform the Laguerre differential equation

$$-xu'' - (1 - x)u' - nu = f$$

to Sturm–Liouville form.

2.5. In SLP1, the following are restrictions on $u(x)$:

(a) $u \in \mathcal{L}_2(a, b)$;
(b) $u \in \mathcal{D}_L$;
(c) u satisfies two boundary conditions, $B_1(u) = 0$, $B_2(u) = 0$.

Show that these restrictions define a linear manifold $\mathcal{M}_L \subset \mathcal{L}_2(a, b)$.

2.6. Solve the following differential equation:

$$-\frac{d^2g(x,\xi)}{dx^2} = \delta(x - \xi)$$

$$g(0,\xi) = \frac{dg(1,\xi)}{dx} = 0$$

2.7. Verify that (2.93) satisfies the requirements for the Green's function given in (2.83)–(2.87).

2.8. Solve the following differential equation:

$$-\frac{d^2 g(x,\xi)}{dx^2} = \delta(x-\xi)$$

$$\frac{dg(0,\xi)}{dx} = \frac{dg(L,\xi)}{dx} = 0$$

2.9. Consider the following differential equation:

$$-u'' = f(x), \qquad x \in (0, L)$$

with $f(x)$ real and with boundary conditions

$$u(0) = \alpha, \qquad \alpha \in \mathbf{R}$$
$$u'(L) = 0$$

(a) Show that, if a solution exists, it is unique.
(b) Construct the solution by the Green's function method.

2.10. Given the differential equation $Lu = f$ on the interval $x \in (0, L)$ with $f(x)$ real and with the following boundary conditions:

$$u(0) = \alpha u'(0), \qquad \alpha > 0$$
$$u'(L) = 0$$

Show that this problem is self-adjoint.

2.11. Show that the solution given in (2.111) satisfies the differential equation and the boundary conditions in Example 2.10.

2.12. Solve the following SLP1 differential equation on the interval $x \in (0, L)$:

$$-u'' - k^2 u = f$$
$$u'(0) = 0$$
$$u'(L) = \beta$$

2.13. Consider the following differential equation:

$$-\frac{d^2 u}{dx^2} = f$$

$$u(0) = \frac{du(1)}{dx}$$

$$\frac{du(0)}{dx} = \alpha, \qquad \alpha \in \mathbf{R}$$

where $x \in (0, 1)$ and where f is a real-valued function of x. Solve the differential equation by the Green's function method.

2.14. For SLP2, show that the operator L is not self-adjoint for the case of initial conditions.

2.15. For SLP2, show that the operator L is self-adjoint for periodic conditions, provided $p(a) = p(b)$.

2.16. Repeat Problem 2.12 for the case where $k \in \mathbf{C}$ and $f(x)$ is complex so that the problem becomes SLP2.

2.17. Repeat Problem 2.9 for the case where $f(x)$ is complex so that the problem becomes SLP2.

2.18. Consider the following SLP2 Green's function problem:

$$-\frac{d^2 g}{dx^2} - \lambda g = \delta(x - \xi)$$

with periodic boundary conditions

$$g(0, \xi) = g(2\pi, \xi)$$

$$\frac{dg(0, \xi)}{dx} = \frac{dg(2\pi, \xi)}{dx}$$

Show that the solution is

$$g(x, \xi) = -\frac{\cos\left[\sqrt{\lambda}(|x - \xi| - \pi)\right]}{2\sqrt{\lambda}\sin\sqrt{\lambda}\pi}$$

2.19. Consider the following SLP3 boundary-value problem, defined on the interval $x \in (0, 1)$:

$$(L - \lambda)u = f$$

$$L = -\frac{d^2}{dx^2}$$

where $\lambda \in \mathbf{C}$ and where the following boundary conditions apply:

$$u'(0) - \alpha u(0) = 0, \qquad \alpha \in \mathbf{C}$$

$$u(1) = 0$$

(a) Show that the operator L is not self-adjoint.

(b) Find the Green's function $g(x, \xi)$, the adjoint Green's function $h(x, \xi)$, and the conjugate adjoint Green's function $\bar{h}(x, \xi)$.

(c) Solve the differential equation.

2.20. Consider the following SLP3 boundary-value problem, defined on the interval $x \in (0, \infty)$:

$$-u'' - k^2 u = f$$

$$u'(0) = 0$$

where $k \in \mathbf{C}$ and $\mathrm{Im}(k) < 0$. Obtain the solution in terms of the appropriate Green's function $g(x, \xi)$ by demanding that

$$\lim_{x \to \infty} [u(x)] = 0$$

2.21. Consider the Green's function problem

$$-\frac{d^2 g}{dx^2} - \lambda g = \delta(x - \xi)$$

with boundary conditions

$$g(0, \xi) = g(b, \xi) = 0$$

Obtain the result in (2.172) by solving this problem and then taking the limit as $b \to \infty$.

REFERENCES

[1] Taylor, A.E. (1955), *Advanced Calculus.* Boston: Ginn, 52–53.

[2] Schwartz, L. (1950), *Théorie des Distributions, Vol. I.* Paris: Hermann.

[3] Stakgold, I. (1967), *Boundary Value Problems of Mathematical Physics, Vol. I.* New York: Macmillan, 18–58.

[4] Papoulis, A. (1962), *The Fourier Integral and Its Applications.* New York: McGraw-Hill, 269–282.

[5] Arfken, G. (1985), *Mathematical Methods for Physicists*, 3rd edition. New York: Academic, 497.

[6] Krall, A.M. (1973), *Linear Methods of Applied Analysis.* Reading, MA: Addison-Wesley, 132.

[7] op.cit. Arfken, 498–499.

[8] Felsen, L.B., and N. Marcuvitz (1973), *Radiation and Scattering of Waves.* Englewood Cliffs, NJ: Prentice-Hall, 323–328.

[9] Hellwig, G. (1964), *Differential Operators of Mathematical Physics.* Reading, MA: Addison-Wesley, 39–40.

[10] op.cit. Taylor, 55.

[11] op.cit. Stakgold, 268–269.

[12] Ibid., 297–301.

[13] Stakgold, I. (1979), *Green's Functions and Boundary Value Problems.* New York: Wiley, 438.

[14] Hajmirzaahmad, M., and A.M. Krall (1992), Singular second- order operators: The maximal and minimal operators, and selfadjoint operators in between, *SIAM Review* 34: 614–634.

[15] Friedman, B. (1956), *Principles and Techniques of Applied Mathematics.* New York: Wiley, 230–231.

[16] Harrington, R.F. (1961, reissued 1987), *Time-Harmonic Electromagnetic Fields.* New York: McGraw-Hill, 461–463.

[17] op.cit. Arfken, 605.

[18] Kaper, H.G., M.K. Kwong, and A. Zettl (1986), Characterizations of the Friedrichs extensions of singular Sturm–Liouville expressions, *SIAM J. Math. Anal.* 17: 772–777.

[19] op.cit., Arfken, 622–636.

[20] Hansen, J.E. (Ed.) (1988), *Spherical Near-Field Antenna Measurements.* London: Peter Perigrinus, 314–315.

[21] op.cit. Hellwig, 221–276.

3

The Spectral Representation Method

3.1 INTRODUCTION

In this chapter, we continue our discussion of linear ordinary differential equations of second order. We begin with a discussion of eigenvalues and eigenfunctions. We follow with a description of the method of solution for self-adjoint SLP2 problems in terms of the eigenfunctions. We include a discussion of the determination of the eigenfunctions directly from the Green's function for the problem. This determination leads to a *spectral representation* of the delta function specific to a particular operator and its domain. We next consider problems on unbounded intervals (SLP3). We are able to expand our analysis to produce the appropriate spectral representations for many important unbounded interval problems. We conclude the chapter by emphasizing the connection between solutions by the Green's function method, and by the spectral representation method.

3.2 EIGENFUNCTIONS AND EIGENVALUES

A complex number μ is called an *eigenvalue* of the linear operator L if there exists a nonzero vector v in the domain of L such that

$$Lv = \mu v \tag{3.1}$$

The vector v is called an *eigenfunction* of the linear operator L. We remark that, although an eigenfunction is by definition nonzero, it can be associated with a zero eigenvalue.

So far, it is unclear whether or not there are any eigenfunctions and eigenvalues associated with a specific operator. However, if eigenvalues and eigenfunctions do exist, they have remarkable properties. We first show that if $u_1, u_2, \ldots, u_n$ are eigenfunctions corresponding to different eigenvalues $\lambda_1, \lambda_2, \ldots, \lambda_n$, associated with the operator L, then $\{u_k\}$ is a linearly independent sequence. Our proof is by induction. Let $n = 1$ and examine

$$\alpha_1 u_1 = 0$$

Since u_1 is an eigenfunction, $u_1 \neq 0$, and therefore $\alpha_1 = 0$. We now suppose that the linearly independent assertion is true for $n-1$ and examine

$$\sum_{k=1}^{n} \alpha_k u_k = 0$$

Operating on both sides with $L - \lambda_n$, we obtain

$$0 = (L - \lambda_n) \sum_{k=1}^{n} \alpha_k u_k = \sum_{k=1}^{n-1} \alpha_k (\lambda_k - \lambda_n) u_k$$

Since the $n - 1$ length sequence has been supposed independent,

$$\alpha_k(\lambda_k - \lambda_n) = 0, \qquad k = 1, 2, \ldots, n-1$$

which implies that

$$\alpha_k = 0, \qquad k = 1, 2, \ldots, n-1$$

The expression

$$\sum_{k=1}^{n} \alpha_k u_k = 0$$

then reduces to

$$\alpha_n u_n = 0$$

which implies that $\alpha_n = 0$. What we have shown is that if the $(n-1)$-length sequence is independent, then the n-length sequence is independent. We have established the result for $n = 1$, and therefore it must be true for $n = 2$. By induction, it must therefore be true for arbitrary n. Further, since n is arbitrary, the countably infinite sequence $u_1, u_2, \ldots$ is linearly independent.

In the above proof of linear independence of the eigenfunctions, we assumed nothing about the operator L except linearity. If the linear operator L is self-adjoint, however, we may show that its eigenvalues are real. Indeed, let μ be an eigenvalue associated with the eigenfunction v. Then,

$$\mu\langle v, v\rangle = \langle \mu v, v\rangle = \langle Lv, v\rangle$$

But, since L is self-adjoint,

$$\mu\langle v, v\rangle = \langle v, Lv\rangle = \langle v, \mu v\rangle = \bar{\mu}\langle v, v\rangle$$

Therefore,

$$(\mu - \bar{\mu})\langle v, v\rangle = 0$$

Since $v \neq 0$, $\langle v, v\rangle > 0$ and we must have

$$\mu - \bar{\mu} = 0$$

which implies that $\mu \in \mathbf{R}$.

We next establish that eigenfunctions of a self-adjoint operator corresponding to different eigenvalues are orthogonal. Indeed, let

$$Lu_m = \lambda_m u_m$$

$$Lu_n = \lambda_n u_n$$

where $\lambda_m \neq \lambda_n$ Then,

$$\lambda_m\langle u_m, u_n\rangle = \langle \lambda_m u_m, u_n\rangle = \langle Lu_m, u_n\rangle$$

Since L is self-adjoint,

$$\lambda_m\langle u_m, u_n\rangle = \langle u_m, Lu_n\rangle = \langle u_m, \lambda_n u_n\rangle = \bar{\lambda}_n\langle u_m, u_n\rangle$$

But, we have established that λ_n is real. Therefore,

$$\lambda_m\langle u_m, u_n\rangle = \lambda_n\langle u_m, u_n\rangle$$

and

$$(\lambda_m - \lambda_n)\langle u_m, u_n\rangle = 0$$

Since $\lambda_m \neq \lambda_n$, we must have $\langle u_m, u_n\rangle = 0$.

We next state a central result for Hilbert space $\mathcal{L}_2(a, b)$. It can be shown that the eigenfunctions u_k of a self-adjoint operator form an orthogonal basis in $\mathcal{L}_2(a, b)$. Therefore, any $u \in \mathcal{L}_2(a, b)$ can be expanded:

$$u = \sum_{k=1}^{\infty} \alpha_k u_k$$

The equality is interpreted in the sense that

$$\lim_{n\to\infty} \|u - \sum_{k=1}^{n} \alpha_k u_k\| = 0$$

The proof of this property involves the theory of integral equations and is beyond the scope of this book. The interested reader is referred to the literature [1].

The fact that the eigenfunctions form an orthogonal basis allows us to solve the self-adjoint SLP2 problem in terms of the eigenfunctions u_n of the operator L. We begin by noting that the eigenfunctions can always be normalized, so that we shall assume that they are orthonormal. Therefore, if

$$u = \sum_{n} \alpha_n u_n \tag{3.2}$$

then by (1.58), the Fourier coefficients are given by

$$\alpha_n = \langle u, u_n \rangle \tag{3.3}$$

where the index n runs over all of the eigenfunctions. On the interval $x \in (a, b)$, consider the following SLP2 problem:

$$(L - \lambda)u = f \tag{3.4}$$

with associated boundary conditions

$$B_1(u) = 0 \tag{3.5}$$

$$B_2(u) = 0 \tag{3.6}$$

We assume that the operator L is self-adjoint. We associate with this SLP2 problem the following *eigenproblem*:

$$Lu_n = \lambda_n u_n \tag{3.7}$$

where $\lambda_n \in \mathbf{R}$. We assign to u_n the same boundary conditions as those we have assigned to u in (3.5) and (3.6), viz.

$$B_1(u_n) = 0 \tag{3.8}$$

$$B_2(u_n) = 0 \tag{3.9}$$

The subscript n is an integer indexing the sequences $\{u_n\}$ and $\{\lambda_n\}$. We form the following inner product relation:

$$\langle (L - \lambda)u, u_n \rangle = \langle u, (L - \bar{\lambda})u_n \rangle + J(u, u_n)\Big|_a^b \tag{3.10}$$

From the self-adjoint property, the conjunct J is zero. Substituting (3.4) and (3.7) into (3.10), we obtain

$$\langle f, u_n \rangle = \langle u, (\lambda_n - \bar{\lambda}) u_n \rangle = (\lambda_n - \lambda)\langle u, u_n \rangle = (\lambda_n - \lambda)\alpha_n$$

Therefore,

$$\alpha_n = \frac{\langle f, u_n \rangle}{\lambda_n - \lambda} \tag{3.11}$$

Substituting (3.11) into (3.2) gives the solution in terms of the eigenvalues and eigenfunctions, viz.

$$u = \sum_n \frac{\langle f, u_n \rangle}{\lambda_n - \lambda} u_n \tag{3.12}$$

EXAMPLE 3.1 Using the eigenfunction–eigenvalue method, we wish to solve the following differential equation on $x \in (0, a)$:

$$(L - \lambda)u = f \tag{3.13}$$

where f and λ are complex and where

$$L = -\frac{d^2}{dx^2} \tag{3.14}$$

and

$$u'(0) = u'(a) = 0 \tag{3.15}$$

The problem is of class SLP2. The operator L with the given unmixed boundary conditions is self-adjoint. The associated eigenproblem is given by

$$-u_n'' = \lambda_n u_n \tag{3.16}$$

with

$$u_n'(0) = u_n'(a) = 0 \tag{3.17}$$

The orthonormal solutions to the eigenproblem are given by

$$u_n = \left(\frac{\epsilon_n}{a}\right)^{1/2} \cos\frac{n\pi x}{a}, \qquad n = 0, 1, \ldots \tag{3.18}$$

where

$$\lambda_n = \left(\frac{n\pi}{a}\right)^2 \tag{3.19}$$

and where ϵ_n is *Neumann's number*, given by

$$\epsilon_n = \begin{cases} 1, & n = 0 \\ 2, & n \neq 0 \end{cases} \tag{3.20}$$

The solution in (3.18) can be easily verified by substitution into (3.16). We note that the factor $\sqrt{\epsilon_n/a}$ normalizes the eigenfunctions. Substitution of (3.18) and (3.19) in (3.12) yields

$$u(x) = \sum_{n=0}^{\infty} \frac{\epsilon_n}{a} \frac{\int_0^a f(x') \cos \frac{n\pi x'}{a} dx'}{\left(\frac{n\pi}{a}\right)^2 - \lambda} \cos \frac{n\pi x}{a} \tag{3.21}$$

■

In Example 3.1, we have solved the differential equation in (3.13)–(3.15) by the eigenfunction–eigenvalue method. The method is also called the *spectral representation method.* The reader should compare the solution in Example 3.1 to the solution by Green's function methods, given in Example 2.13. It appears that the Green's function method might always be preferred since the spectral representation method contains a summation that must be performed before the mathematics can be reduced to a numerical answer. There are, however, many reasons why the spectral representation is important. First, consider (3.21) in a slightly different form, viz.

$$u(x) = \sum_{n=0}^{\infty} A_n \cos \frac{n\pi x}{a} \tag{3.22}$$

where

$$A_n = \frac{\epsilon_n}{a\left[\left(\frac{n\pi}{a}\right)^2 - \lambda\right]} \int_0^a f(x') \cos(\frac{n\pi x'}{a}) dx' \tag{3.23}$$

We interpret (3.22) as a Fourier sum over the *natural modes* of the system with *modal coefficients* A_n. We note that it is the interaction between the forcing function $f(x)$ and the modes in (3.23) that determines the modal coefficients. Second, in dealing with multidimensional problems in later chapters, we shall encounter partial differential equations whose solutions are rarely expressible in terms of closed-form Green's functions. However, proper combination of the spectral representation and Green's function methods results in a powerful tool to solve many partial differential equations appearing in electromagnetic problems.

There is a variation on the procedure used to produce the result in (3.12). This variation results in a more direct approach to the solution of differential equations by the eigenfunction–eigenvalue method. In addition, the variation is very useful in the solution to the partial differential equations considered in later chapters. We proceed as follows. Suppose L is self-adjoint with associated orthonormal eigenfunctions u_n. Then,

$$u = \sum_n \alpha_n u_n \tag{3.24}$$

$$\alpha_n = \langle u, u_n \rangle \tag{3.25}$$

We say that (3.25) is a transformation of the function $u(x) \in \mathcal{L}_2(a, b)$ into coefficients α_n. Conversely, (3.24) is the inverse transformation of the coefficients α_n into the function $u(x)$. We represent this transformation relationship by $u(x) \Longleftrightarrow \alpha_n$. We now show that if L is self-adjoint and

$$u \Longleftrightarrow \alpha_n \tag{3.26}$$

then,

$$Lu \Longleftrightarrow \lambda_n \alpha_n \tag{3.27}$$

Indeed, forming the transformation defined in (3.25), we have

$$\langle Lu, u_n \rangle = \langle u, Lu_n \rangle = \lambda_n \langle u, u_n \rangle = \lambda_n \alpha_n \tag{3.28}$$

Having established the basic result in (3.27), we reconsider the original problem stated in (3.4)–(3.6), which we repeat here for convenience, viz.

$$(L - \lambda)u = f \tag{3.29}$$

with associated boundary conditions

$$B_1(u) = 0 \tag{3.30}$$

$$B_2(u) = 0 \tag{3.31}$$

where we assume that L is self-adjoint. If u_n are eigenfunctions and λ_n are eigenvalues of L, then

$$u \Longleftrightarrow \alpha_n \tag{3.32}$$

and

$$Lu \Longleftrightarrow \lambda_n \alpha_n \tag{3.33}$$

If we define

$$f(x) \Longleftrightarrow \beta_n \tag{3.34}$$

then (3.29) transforms into

$$(\lambda_n - \lambda)\alpha_n = \beta_n \tag{3.35}$$

Solving for α_n, we obtain

$$\alpha_n = \frac{\beta_n}{\lambda_n - \lambda} = \frac{\langle f, u_n \rangle}{\lambda_n - \lambda} \tag{3.36}$$

Substitution of this result into (3.24) yields (3.12).

3.3 SPECTRAL REPRESENTATIONS FOR SLP1 AND SLP2

In Example 3.1, we assumed that we had somehow obtained the eigenfunctions given by (3.18). In this section, we shall present a method of obtaining the eigenfunctions and eigenvalues of a self-adjoint operator directly from the Green's function for the problem. Since SLP1 is a special case contained in SLP2, we shall confine our attention to SLP2.

Note that the solution in (3.12) in terms of the eigenfunctions and eigenvalues is parametrically dependent on λ, viz.

$$u(x, \lambda) = -\sum_n \frac{\langle f, u_n \rangle}{\lambda - \lambda_n} u_n$$

Consider

$$\oint_{C_R} u(x, \lambda) d\lambda$$

where C_R is a circle of radius R centered at the origin in the complex λ-plane (Fig. 3-1). We have

$$\oint_{C_R} u(x, \lambda) d\lambda = -\sum_n \langle f, u_n \rangle u_n \oint_{C_R} \frac{d\lambda}{\lambda - \lambda_n}$$

where the sum is over those eigenvalues λ_n contained within the circle. The singularities of the integrand are simple poles with residue of unity at all $\lambda = \lambda_n$ within the contour. We note that since L is self-adjoint, the poles must lie on the real axis in the λ-plane. Taking the limit as $R \to \infty$, we enclose all of the singularities and obtain by the *Residue Theorem* [2]

$$\lim_{R \to \infty} \oint_{C_R} u(x, \lambda) d\lambda = -2\pi i \sum_n \langle f, u_n \rangle u_n \tag{3.37}$$

where the sum is now over all of the eigenfunctions. The summation is simply the Fourier expansion of the forcing function in terms of the eigenfunctions. Therefore, we find that

$$\frac{1}{2\pi i} \oint_C u(x, \lambda) d\lambda = -f(x) \tag{3.38}$$

where C is the contour at infinity obtained in the limiting operation in (3.37). There is an important special case to the result in (3.38). We note that the general forcing function $f(x)$ produces the response $u(x, \lambda)$.

Therefore, the specific forcing function $\delta(x-\xi)/w(x)$ must produce the Green's function $g(x, \xi, \lambda)$. We therefore obtain

$$\frac{1}{2\pi i}\oint_C g(x, \xi, \lambda)d\lambda = -\frac{\delta(x-\xi)}{w(x)} \tag{3.39}$$

The solution to the contour integral in (3.39) for a specific Green's function associated with a specific operator L and boundary conditions is called the *spectral representation of the delta function* [3] for the operator L. We shall demonstrate the utility of this result in an example.

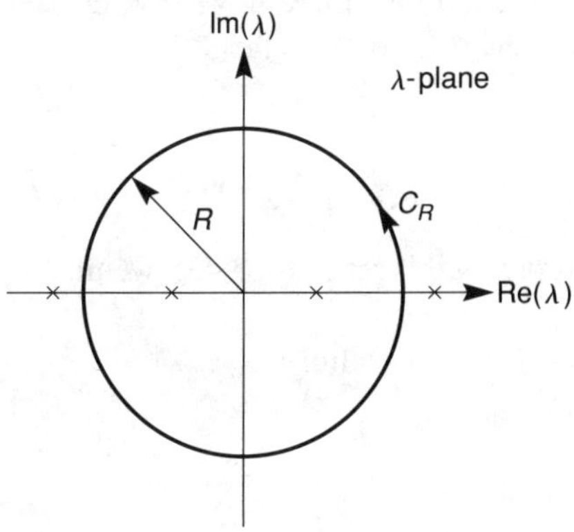

Fig. 3-1 Circular contour of radius R centered at the origin in the complex λ-plane. The $\times$ indicate possible simple pole locations at $\lambda = \lambda_n$.

EXAMPLE 3.2 Consider the following operator defined on $x \in (0, a)$:

$$L = -\frac{d^2}{dx^2} \tag{3.40}$$

with boundary conditions

$$u'(0) = u'(a) = 0 \tag{3.41}$$

The Green's function associated with L_λ with these boundary conditions has been previously derived in Example 2.13. We repeat it here for convenience, viz.

$$g(x, \xi, \lambda) = -\frac{1}{\sqrt{\lambda}\sin\sqrt{\lambda}a}\begin{cases}\cos\sqrt{\lambda}x\cos\sqrt{\lambda}(a-\xi), & x < \xi\\ \cos\sqrt{\lambda}\xi\cos\sqrt{\lambda}(a-x), & x > \xi\end{cases} \tag{3.42}$$

First, consider the case $x < \xi$, so that

$$g(x, \xi, \lambda) = -\frac{\cos\sqrt{\lambda}x\cos\sqrt{\lambda}(a-\xi)}{\sqrt{\lambda}\sin\sqrt{\lambda}a} \tag{3.43}$$

Substitution into (3.39) gives

$$\delta(x-\xi) = \frac{1}{2\pi i}\oint_C \frac{\cos\sqrt{\lambda}x\cos\sqrt{\lambda}(a-\xi)}{\sqrt{\lambda}\sin\sqrt{\lambda}a}d\lambda \tag{3.44}$$

In order to solve this closed contour integral by the residue theorem, we first investigate the singularities of the integrand. Since $\sqrt{\lambda}$ is a multiple-valued function, we might expect that the integrand contains a branch cut with a branch point at $\lambda = 0$. We may show, however, that although $\sqrt{\lambda}$ has a branch cut, $g(x, \xi, \lambda)$ does not. Indeed, define

$$\lambda = |\lambda| e^{i\phi}, \qquad 2\pi > \phi > 0 \tag{3.45}$$

so that

$$\sqrt{\lambda} = |\lambda|^{1/2} e^{i\phi/2}, \quad \pi > \frac{\phi}{2} > 0 \tag{3.46}$$

This definition of λ results in a branch cut in $\sqrt{\lambda}$ along the positive-real axis in the λ-plane (Fig. 3-2). In fact,

$$\lim_{\phi \to 0} \sqrt{\lambda} = |\lambda|^{1/2}$$

$$\lim_{\phi \to 2\pi} \sqrt{\lambda} = -|\lambda|^{1/2}$$

Applying this result to (3.43), we find that

$$\lim_{\phi \to 0} g(x, \xi, \lambda) = -\frac{\cos |\lambda|^{1/2} x \cos |\lambda|^{1/2}(a - \xi)}{|\lambda|^{1/2} \sin |\lambda|^{1/2} a}$$

$$\lim_{\phi \to 2\pi} g(x, \xi, \lambda) = -\frac{\cos(-|\lambda|^{1/2} x) \cos[-|\lambda|^{1/2}(a - \xi)]}{-|\lambda|^{1/2} \sin(-|\lambda|^{1/2} a)}$$

Some minor algebraic manipulation shows that

$$\lim_{\phi \to 0} g(x, \xi, \lambda) = \lim_{\phi \to 2\pi} g(x, \xi, \lambda)$$

We conclude that, although there is a branch cut in $\sqrt{\lambda}$ along the positive-real axis, the Green's function is continuous there. We next consider the possible location and order of poles of $g(x, \xi, \lambda)$. Since the numerator of the Green's function in (3.43) is finite throughout the complex λ-plane, it is sufficient to search for poles caused by the denominator. Let

$$f(\lambda) = \frac{1}{\sqrt{\lambda} \sin \sqrt{\lambda} a} = \frac{p(\lambda)}{q(\lambda)}$$

where

$$p(\lambda) = 1$$

$$q(\lambda) = \sqrt{\lambda} \sin \sqrt{\lambda} a$$

By a well-known theorem of complex analysis [4], $f(\lambda)$ has a simple pole at $\lambda = \lambda_n$ if $p(\lambda_n) \neq 0$, $q(\lambda_n) = 0$, and $q'(\lambda_n) \neq 0$. In addition, the residue at the simple pole location is given by

$$\text{Res}\{f(\lambda); \lambda_n\} = \frac{p(\lambda_n)}{q'(\lambda_n)}$$

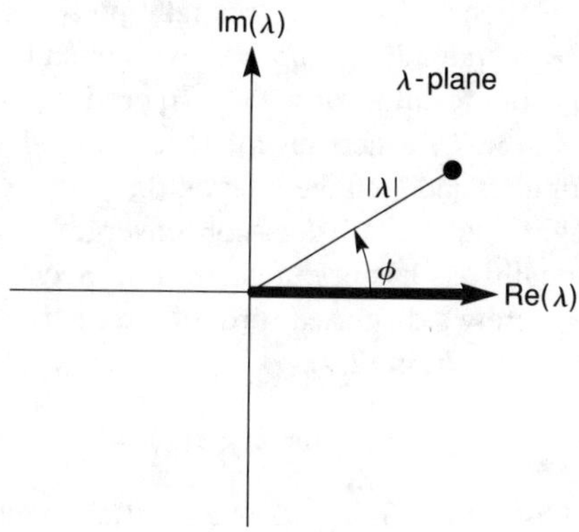

Fig. 3-2 Polar representation of λ in the complex λ-plane showing branch cut for $\sqrt{\lambda}$ along positive-real axis (thick line).

We note that $p(\lambda) \neq 0$ anywhere, and that $q(\lambda) = 0$ wherever $\sqrt{\lambda}a = 0, \pm\pi, \pm 2\pi, \ldots$. We conclude that $q(\lambda) = 0$ whenever

$$\lambda_n = \left(\frac{n\pi}{a}\right)^2, \qquad n = 0, 1, 2, \ldots$$

Differentiating the denominator, we find that

$$q'(\lambda) = \frac{a}{2}\left(\cos\sqrt{\lambda}a + \frac{\sin\sqrt{\lambda}a}{\sqrt{\lambda}a}\right)$$

from which

$$q'(\lambda_n) = (-1)^n\left(\frac{a}{\epsilon_n}\right), \qquad n = 0, 1, \ldots$$

where ϵ_n is Neumann's number, defined in (3.20). We have shown that the singularities at λ_n are simple poles. For the residues, we have

$$\text{Res}\{f(\lambda); \lambda_n\} = (-1)^n \frac{\epsilon_n}{a}$$

Returning to (3.44), we now have

$$\delta(x-\xi) = \sum_{n=0}^{\infty} \cos\frac{n\pi x}{a} \cos\frac{n\pi}{a}(a-\xi)\text{Res}\{f(\lambda); \lambda_n\}$$

$$[8pt] = \sum_{n=0}^{\infty}(-1)^n \frac{\epsilon_n}{a} \cos\frac{n\pi x}{a} \cos\frac{n\pi}{a}(a-\xi)$$

and finally,

$$\delta(x-\xi) = \sum_{n=0}^{\infty} \frac{\epsilon_n}{a} \cos\frac{n\pi x}{a} \cos\frac{n\pi\xi}{a} \tag{3.47}$$

Equation (3.47) is the spectral representation of the delta function associated with the operator $L = -d^2/dx^2$ with boundary conditions $u'(0) = u'(a) = 0$. Recall that our result is for $x < \xi$. To produce the result for $x > \xi$, we interchange x and ξ in (3.47). Since this interchange produces no change in the result, (3.47) holds for all x and ξ on the interval $(0, a)$. The reader is cautioned that the series on the right side of (3.47) does not converge. It is, however, an extremely useful symbolic equality, as has been discussed in Section 2.2. Indeed, we may show that (3.47) is merely a disguised form of the Fourier cosine series. For any $s(x) \in \mathcal{L}_2(0, a)$, we have, from (2.9),

$$s(x) = \int_0^a \delta(x - \xi)s(\xi)d\xi \tag{3.48}$$

Substituting (3.47), we obtain, after an interchange of integration and summation,

$$s(x) = \sum_{n=0}^{\infty} \alpha_n \sqrt{\frac{\epsilon_n}{a}} \cos \frac{n\pi x}{a}$$

where

$$\alpha_n = \int_0^a s(\xi)\sqrt{\frac{\epsilon_n}{a}} \cos \frac{n\pi\xi}{a} d\xi$$

We note that we have produced the eigenfunctions and eigenvalues inferred in (3.18) and (3.19) directly from the Green's function associated with the operator L and its boundary conditions.

■

In the example spectral representation in (3.47), each term in the sum consists of the product of the orthonormal eigenfunction as a function of x with the same orthonormal eigenfunction as a function of ξ. We may show that this result can be generalized to all self-adjoint operators on SLP2. Indeed, if the forcing function in (3.12) is the delta function, we have

$$g(x, \xi, \lambda) = \sum_n \frac{\langle \frac{\delta(x'-\xi)}{w(x')}, u_n(x') \rangle_{x'}}{\lambda_n - \lambda} u_n(x)$$

where, as indicated, the inner product is with respect to x'. Performing the integration gives

$$g(x, \xi) = \sum_n \frac{u_n(x)\overline{u_n}(\xi)}{\lambda_n - \lambda} \tag{3.49}$$

This form of the Green's function is called the bilinear series form [5]. Substitution of (3.49) into (3.39) gives

$$\frac{\delta(x - \xi)}{w(x)} = -\frac{1}{2\pi i} \sum_n u_n(x)\overline{u_n}(\xi) \oint_C \frac{d\lambda}{\lambda_n - \lambda}$$

or

$$\frac{\delta(x-\xi)}{w(x)} = \sum_n u_n(x)\overline{u_n}(\xi) \tag{3.50}$$

Equation (3.50) gives the general spectral representation for self-adjoint operators on SLP2. The procedure for solving self-adjoint SLP2 problems by the spectral representation method can now be summarized as follows:

1. For a given self-adjoint operator L and given boundary conditions, solve the Green's function problem $L_\lambda g(x, \xi, \lambda) = \delta(x-\xi)/w(x)$.
2. Substitute the Green's function $g(x, \xi, \lambda)$ into (3.39) and solve for the spectral representation of the delta function. The resulting form should appear as in (3.50).
3. Substitute the normalized eigenfunctions and their eigenvalues, obtained in the spectral representation, into (3.12) to produce the solution $u(x)$.

In addition to Example 3.2, we have included in the problems several common examples to illustrate the method.

3.4 SPECTRAL REPRESENTATIONS FOR SLP3

The spectral representation of the delta function takes on a different character when the interval along the real axis becomes unbounded. Consider the problem in Example 3.2, defined on the interval $x \in (0, a)$. The Green's function $g(x, \xi, \lambda)$ was shown to have simple poles at $\lambda = (n\pi/a)^2$. Note that as a becomes larger, the poles become closer together. In the limit as $a \to \infty$, the poles become arbitrarily close. This behavior leads us to inquire into the form of the singularity along the positive-real axis in the λ-plane in this limiting case. We shall illustrate the obtaining of the spectral representation with an example.

EXAMPLE 3.3 Consider the following SLP3 differential equation:

$$-u'' - \lambda u = f, \qquad \lambda \in \mathbf{C} \tag{3.51}$$

$$u(0) = 0 \tag{3.52}$$

This problem is in the limit point case as $x \to \infty$. We therefore assign the limit condition

$$\lim_{x\to\infty} u(x) = 0 \tag{3.53}$$

The associated Green's function problem is

$$-\frac{d^2g}{dx^2} - \lambda g = \delta(x - \xi)$$
$$g(0, \xi) = 0$$
$$\lim_{x\to\infty} g(x, \xi) = 0$$

We have previously obtained this Green's function in (2.172) and repeat it here for convenience, viz.

$$g(x, \xi) = \frac{1}{\sqrt{\lambda}} \begin{cases} e^{-i\sqrt{\lambda}\xi} \sin \sqrt{\lambda}x, & x < \xi \\ e^{-i\sqrt{\lambda}x} \sin \sqrt{\lambda}\xi, & x > \xi \end{cases} \tag{3.54}$$

where

$$\text{Im}\sqrt{\lambda} < 0 \tag{3.55}$$

To produce the restriction in (3.55), we define

$$\lambda = |\lambda| e^{i\phi}, \quad 2\pi < \phi < 4\pi \tag{3.56}$$

so that

$$\sqrt{\lambda} = |\lambda|^{1/2} e^{i\phi/2}, \quad \pi < \frac{\phi}{2} < 2\pi \tag{3.57}$$

The angular definition in (3.56) defines a *Riemann sheet* of the λ-plane. The angular restriction in (3.57) indicates that $\text{Im}\sqrt{\lambda} < 0$ everywhere on this sheet. We shall refer to this sheet as the *proper Riemann sheet*. We note that once (3.55) has been invoked, any result with $\text{Im}\sqrt{\lambda} > 0$ would violate the requirement for the proper Riemann sheet. The definition of λ in (3.56) results in a branch cut in $\sqrt{\lambda}$ along the positive-real axis in the λ-plane (Fig. 3-2). In this case,

$$\lim_{\phi\to 2\pi} \sqrt{\lambda} = -|\lambda|^{1/2}$$
$$\lim_{\phi\to 4\pi} \sqrt{\lambda} = |\lambda|^{1/2}$$

Unlike the situation in Example 3.2, however, the branch cut in $\sqrt{\lambda}$ produces a branch cut in $g(x, \xi, \lambda)$ along the positive-real axis. Indeed, for $x < \xi$,

$$\lim_{\phi\to 2\pi} g(x, \xi, \lambda) = \frac{e^{i|\lambda|^{1/2}\xi} \sin |\lambda|^{1/2}x}{|\lambda|^{1/2}}$$

$$\lim_{\phi\to 4\pi} g(x, \xi, \lambda) = \frac{e^{-i|\lambda|^{1/2}\xi} \sin |\lambda|^{1/2}x}{|\lambda|^{1/2}}$$

Since the exponential changes sign, $g(x, \xi, \lambda)$ is discontinuous across the positive-real axis. The result for $x > \xi$ is the same, except that x and ξ are interchanged.

We next produce the spectral representation by considering the closed contour in the complex λ-plane shown in Fig. 3-3. Since the contour excludes the branch cut and since $g(x, \xi, \lambda)$ has no other singularities, Cauchy's theorem gives

$$\oint_{C_R+C_1+C_\rho+C_2} g(x, \xi, \lambda)d\lambda = 0 \tag{3.58}$$

We examine the contributions from the various portions of the contour in the limit as $\rho \to 0$, $R \to \infty$, and $\Gamma \to 0$. Consider C_R. From (3.39), we have

$$\lim_{\Gamma\to 0} \lim_{R\to\infty} \int_{C_R} g(x, \xi, \lambda)d\lambda = -2\pi i\delta(x - \xi) \tag{3.59}$$

where we have assumed that, even in the presence of the branch cut, the integral around the circle of infinite radius centered at the origin produces the delta-function contribution. Consider C_ρ. Letting $\lambda = \rho \exp(i\phi)$, we have

$$\int_{C_\rho} g(x, \xi, \lambda)d\lambda = i\rho \int_{4\pi-\Gamma}^{2\pi+\Gamma} \frac{\sin(\rho^{1/2}e^{i\phi/2}x)}{\rho^{1/2}e^{i\phi/2}} e^{-i(\rho^{1/2}e^{i\phi/2}\xi)} e^{i\phi} d\phi$$

Since the integrand on the right side is bounded as $\rho \to 0$ and $\Gamma \to 0$, the integral is bounded and

$$\lim_{\Gamma\to 0} \lim_{\rho\to 0} \int_{C_\rho} g(x, \xi, \lambda)d\lambda = 0 \tag{3.60}$$

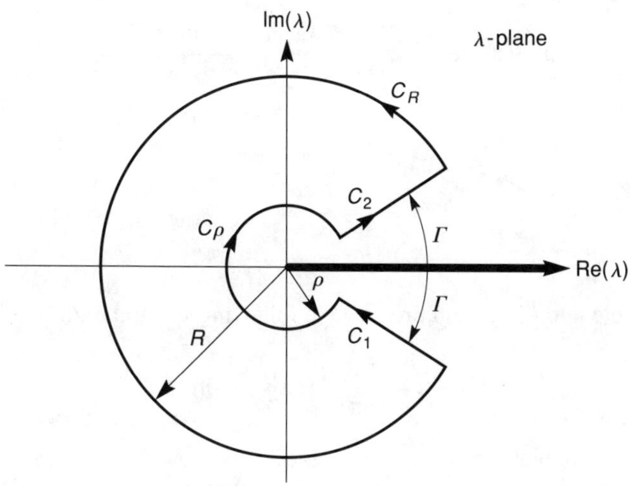

Fig. 3-3 Contour for evaluation of the spectral representation for Example 3.3.

Consider the integral along C_1 and C_2. On C_1, let

$$\lambda = re^{i(4\pi-\Gamma)}$$

On C_2, let

$$\lambda = re^{i(2\pi+\Gamma)}$$

We obtain

$$\int_{C_1+C_2} g(x,\xi,\lambda)d\lambda = \int_R^\rho \frac{\sin[r^{1/2}e^{i(2\pi-\Gamma/2)}x]}{r^{1/2}e^{i(2\pi-\Gamma/2)}} e^{-i[r^{1/2}e^{i(2\pi-\Gamma/2)}\xi]}e^{i(4\pi-\Gamma)}dr + \int_\rho^R \frac{\sin[r^{1/2}e^{i(\pi+\Gamma/2)}x]}{r^{1/2}e^{i(\pi+\Gamma/2)}} e^{-i[r^{1/2}e^{i(\pi+\Gamma/2)}\xi]}e^{i(2\pi+\Gamma)}dr \tag{3.61}$$

Taking the limits, we have

$$\lim_{\rho\to 0}\lim_{R\to\infty}\lim_{\Gamma\to 0}\int_{C_1+C_2} g(x,\xi,\lambda)d\lambda = \int_\infty^0 \frac{\sin r^{1/2}x}{r^{1/2}}e^{-ir^{1/2}\xi}dr + \int_0^\infty \frac{\sin r^{1/2}x}{r^{1/2}}e^{ir^{1/2}\xi}dr = 2i\int_0^\infty \frac{\sin(r^{1/2}x)\sin(r^{1/2}\xi)}{r^{1/2}}dr \tag{3.62}$$

Combining the results in (3.58)–(3.62), we obtain

$$-2\pi i\delta(x-\xi) + 2i\int_0^\infty \frac{\sin(r^{1/2}x)\sin(r^{1/2}\xi)}{r^{1/2}}dr = 0 \tag{3.63}$$

We let

$$k = r^{1/2}$$

so that

$$dk = \frac{dr}{2r^{1/2}}$$

We substitute into (3.63) and produce the following spectral representation:

$$\delta(x-\xi) = \frac{2}{\pi}\int_0^\infty \sin kx \sin k\xi dk \tag{3.64}$$

In a similar manner to that in Example 3.2, we now show that (3.64) is merely a disguised form of the Fourier sine transform. Indeed, for $f(x) \in \mathcal{L}_2(0,\infty)$, we have

$$f(x) = \int_0^\infty \delta(x-\xi)f(\xi)d\xi$$

Substituting (3.64), we obtain, after an interchange of integrations,

$$f(x) = \frac{2}{\pi} \int_0^{\infty} F(k) \sin kx dk \tag{3.65}$$

where

$$F(k) = \int_0^{\infty} f(\xi) \sin k\xi d\xi \tag{3.66}$$

We indicate the Fourier sine transform relationship symbolically by

$$f(x) \Longleftrightarrow F(k) \tag{3.67}$$

We now return to the solution to the differential equation considered in (3.51)–(3.53), repeated here for convenience. Consider $\mathcal{L}_2(0, \infty)$ with inner product

$$\langle u, v \rangle = \int_0^{\infty} u(x)v(x)dx \tag{3.68}$$

Consider

$$-u'' - \lambda u = f, \qquad \lambda \in \mathbf{C} \tag{3.69}$$

$$u(0) = 0 \tag{3.70}$$

$$\lim_{x \to \infty} u(x) = 0 \tag{3.71}$$

From Example 2.18 and the discussion following, this problem is self-adjoint. Using the results in (3.66) and (3.67), we expand $u(x)$ as follows:

$$u(x) = \frac{2}{\pi} \int_0^{\infty} U(k) \sin kx dk \tag{3.72}$$

where

$$U(k) = \langle u, \sin kx \rangle = \int_0^{\infty} u(\xi) \sin k\xi d\xi \tag{3.73}$$

Expression (3.72) is the equivalent to (3.24), except in this case we have an integral, rather than a sum. We note that $\sin kx$ plays the role that the eigenfunction u_n plays in (3.24). In (3.73), $U(k)$ is similar to the Fourier coefficient in (3.25), where $\sin kx$ plays the same role as the eigenfunction u_n in (3.25). To further investigate the similarity to the eigenfunction u_n, we note that

$$-\frac{d^2 \sin kx}{dx^2} = k^2 \sin kx \tag{3.74}$$

so that $\sin kx$ appears to be an eigenfunction of the self-adjoint operator $-d^2/dx^2$ with eigenvalue k^2. However, $\sin kx$ is not in $\mathcal{L}_2(0, \infty)$, and therefore cannot be an eigenfunction. We shall adopt the notation of Friedman [6] and call $\sin kx$ an *improper eigenfunction* with *improper eigenvalue* k^2. Fortunately, the procedure we adopted in (3.24)–(3.36) for solving differential equations by the eigenfunction–eigenvalue method can be extended to apply to improper eigenfunctions. We begin by showing that if

$$u \Longleftrightarrow U \tag{3.75}$$

then

$$-\frac{d^2u}{dx^2} \Longleftrightarrow k^2U \tag{3.76}$$

Indeed, following (3.28), we form

$$\begin{aligned}\langle -\frac{d^2u}{dx^2}, \sin kx\rangle &= \langle u, -\frac{d^2 \sin kx}{dx^2}\rangle + J(u, \sin kx)\Big|_0^\infty \\ &= k^2U + \left(-\frac{du}{dx}\sin kx + ku\cos kx\right)\Bigg|_0^\infty \end{aligned} \tag{3.77}$$

Using the conditions in (3.70) and (3.71), we have

$$\langle -\frac{d^2u}{dx^2}, \sin kx\rangle = k^2U - \lim_{x\to\infty}\frac{du}{dx}\sin kx \tag{3.78}$$

We note that the improper eigenfunction $\sin kx$ does not vanish in the limit as $x \to \infty$. This behavior is contrary to what we found when dealing with eigenfunctions on finite intervals. Fortunately, in electromagnetic problems, when we have

$$\lim_{x\to\infty} u(x) = 0$$

then also

$$\lim_{x\to\infty}\frac{du(x)}{dx} = 0$$

For example, if u is a component of the electric field, then $\partial u/\partial x$ is a component of the magnetic field; if the E-field vanishes at infinity, then the H-field also vanishes. Therefore, in the usual cases in electromagnetics, we obtain

$$\langle -\frac{d^2u}{dx^2}, \sin kx\rangle = k^2U \tag{3.79}$$

which establishes (3.76). As a footnote, we remark that there are mathematical theorems that generalize this result to classes of functions possessing certain continuity and absolute integrability properties. The interested reader is referred to [7].

We now are able to solve the original differential equation in (3.69) using the spectral representation. Taking the Fourier sine transform of both sides of (3.69), we obtain

$$(k^2 - \lambda)U(k) = F(k) \tag{3.80}$$

Dividing both sides by $(k^2 - \lambda)$ and taking the inverse transform, we obtain

$$u(x) = \frac{2}{\pi} \int_0^\infty \frac{F(k)}{k^2 - \lambda} \sin kx dk \tag{3.81}$$

■

In the above example, we have indicated explicitly the various limiting operations involved in evaluating the integral of the Green's function around the closed contour indicated in Fig. 3-3. In subsequent discussions, we shall, whenever appropriate, simplify the contour (Fig. 3-4) such that the limiting operations have already taken place. The contour in Fig. 3-4 is to be interpreted as follows. The contour segments C_1 and C_2 are straight lines that are the result of limits as we approach the branch cut from below and above, respectively. By Cauchy's Theorem, if there are no singularities inside the contour, then the integral along C_R is the negative of the integral along $C_1 + C_2$. Therefore, we have

$$\int_{C_1+C_2} g(x, \xi, \lambda) d\lambda = -\int_{C_R} g(x, \xi, \lambda) d\lambda = \frac{2\pi i \delta(x - \xi)}{w(x)} \tag{3.82}$$

In (3.82), we have assumed that there is no contribution obtained from the integral along C_ρ (Fig. 3-3) in the limit as $\rho \to 0$. We use this abbreviated method in the following example.

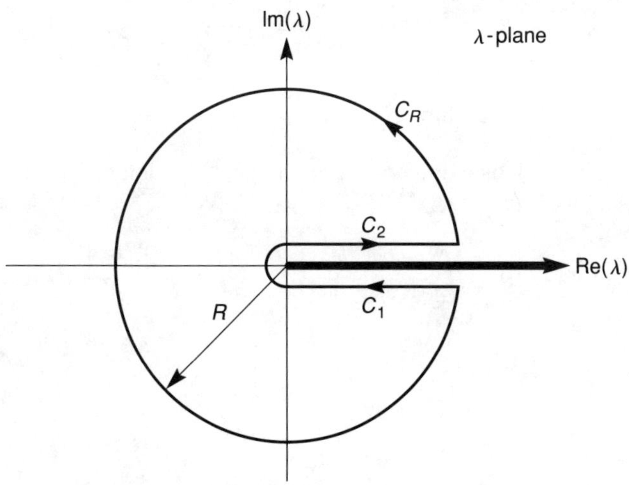

Fig. 3-4 Simplified contour for evaluation of the spectral representation of the delta function for SLP3 problems.

EXAMPLE 3.4 Consider Hilbert space $\mathcal{L}_2(-\infty, \infty)$ with inner product

$$\langle u, v\rangle = \int_{-\infty}^{\infty} u(x)\overline{v}(x)dx \tag{3.83}$$

We seek the spectral representation for the self-adjoint operator

$$L = -d^2/dx^2 \tag{3.84}$$

with limiting conditions

$$\lim_{x\to-\infty} u(x) = \lim_{x\to\infty} u(x) = 0 \tag{3.85}$$

In Example 2.20, we considered the following Green's function problem:

$$-\frac{d^2 g}{dx^2} - \lambda g = \delta(x-\xi), \qquad \mathrm{Im}\sqrt{\lambda} < 0 \tag{3.86}$$

$$\lim_{x\to-\infty} g(x,\xi) = \lim_{x\to\infty} g(x,\xi) = 0 \tag{3.87}$$

The solution to this problem was given in (2.175), viz.

$$g(x,\xi) = \frac{e^{-i\sqrt{\lambda}|x-\xi|}}{2i\sqrt{\lambda}} \tag{3.88}$$

The singularities of $g(x,\xi)$ involve the branch cut associated with $\sqrt{\lambda}$. We define this branch cut using (3.56) and (3.57), and find for $x > \xi$

$$\lim_{\phi\to 2\pi} g(x,\xi,\lambda) = \frac{e^{i|\lambda|^{1/2}(x-\xi)}}{-2i|\lambda|^{1/2}} \tag{3.89}$$

$$\lim_{\phi\to 4\pi} g(x,\xi,\lambda) = \frac{e^{-i|\lambda|^{1/2}(x-\xi)}}{2i|\lambda|^{1/2}} \tag{3.90}$$

Therefore (Fig. 3-4), there is a branch cut in $g(x,\xi,\lambda)$ along the positive-real axis. Using (3.82), we obtain

$$2\pi i\delta(x-\xi) = \int_{\infty}^{0} \frac{e^{-i|\lambda|^{1/2}(x-\xi)}}{2i|\lambda|^{1/2}} d\lambda - \int_{0}^{\infty} \frac{e^{i|\lambda|^{1/2}(x-\xi)}}{2i|\lambda|^{1/2}} d\lambda \tag{3.91}$$

We let

$$k = |\lambda|^{1/2} \tag{3.92}$$

and find that

$$2\pi\delta(x-\xi) = -\int_{\infty}^{0} e^{-ik(x-\xi)} dk + \int_{0}^{\infty} e^{ik(x-\xi)} dk \tag{3.93}$$

Replacing k by $-k$ in the first integral gives the final result, viz.

$$\delta(x-\xi) = \frac{1}{2\pi}\int_{-\infty}^{\infty} e^{ik(x-\xi)}dk \tag{3.94}$$

We have obtained this result for the case $x > \xi$. However, this restriction can be removed. Indeed, to obtain the case $x < \xi$, we merely interchange x and ξ in (3.94). However,

$$\delta(x-\xi) = \delta(\xi - x)$$

which means that we can again reverse the interchange and reclaim the result in (3.94).

We next use the spectral representation in (3.94) to produce the *Fourier transform*. We write

$$u(x) = \int_{-\infty}^{\infty} u(\xi)\delta(x-\xi)d\xi \tag{3.95}$$

Substitution of (3.94) followed by a change in the order of integration yields

$$u(x) = \frac{1}{2\pi}\int_{-\infty}^{\infty} U(k)e^{ikx}dk \tag{3.96}$$

where

$$U(k) = \int_{-\infty}^{\infty} u(x)e^{-ikx}dx = \langle u, e^{ikx}\rangle \tag{3.97}$$

In (3.96) and (3.97), we identify $\exp(ikx)$ as an improper eigenfunction with improper eigenvalue k^2. We indicate the Fourier transform relationship symbolically by

$$u(x) \Longleftrightarrow U(k) \tag{3.98}$$

We may use the Fourier transform to solve the following differential equation by the spectral representation method:

$$-u'' - \lambda u = f, \qquad \lambda \in \mathbf{C} \tag{3.99}$$

$$\lim_{x\to-\infty} u(x) = 0 \tag{3.100}$$

$$\lim_{x\to\infty} u(x) = 0 \tag{3.101}$$

We begin by showing that if

$$u(x) \Longleftrightarrow U(k)$$

then

$$-u''(x) \Longleftrightarrow k^2 U(k) \tag{3.102}$$

Indeed,

$$\begin{aligned}\langle -u'', e^{ikx}\rangle &= \langle u, -\frac{d^2 e^{ikx}}{dx^2}\rangle + J(u, e^{ikx})\Big|_{-\infty}^{\infty} \\ &= k^2 U(k) + \left(-u'e^{-ikx} - ikue^{-ikx}\right)\Big|_{-\infty}^{\infty}\end{aligned} \tag{3.103}$$

Using the conditions in (3.100) and (3.101), we have

$$\langle -u'', e^{ikx} \rangle = k^2 U(k) - \left(u' e^{-ikx}\right)\Big|_{-\infty}^{\infty}$$

From the discussion associated with (3.78), we know that in the usual cases in electromagnetics, (3.100) and (3.101) imply that

$$\lim_{x \to \pm\infty} u'(x) = 0$$

and therefore,

$$\langle -u'', e^{ikx} \rangle = k^2 U(k) \tag{3.104}$$

which proves (3.102). We now take the Fourier transform of both sides of (3.99) and obtain

$$(k^2 - \lambda) U(k) = F(k)$$

Dividing both sides by $(k^2 - \lambda)$ and taking the inverse Fourier transform, we have

$$u(x) = \frac{1}{2\pi} \int_{-\infty}^{\infty} \frac{F(k)}{k^2 - \lambda} e^{ikx} dk \tag{3.105}$$

We note that the result in (3.102) could also be obtained by twice differentiating (3.96), *provided* that we can interchange differentiation and integration on the right side. Our method of proof provides a justification of this interchange in this case. ■

We next provide two examples leading to solutions involving Bessel functions. These examples will be useful in problems in cylindrical coordinates to be considered in later chapters.

EXAMPLE 3.5 Consider the following differential equation on $x \in (0, \infty)$:

$$(L - \lambda) u = f \tag{3.106}$$

where

$$L = -\frac{1}{x}\left[\frac{d}{dx}\left(x \frac{d}{dx}\right)\right] \tag{3.107}$$

From Example 2.21, this problem is in the limit circle case as $x \to 0$ and the limit point case as $x \to \infty$. Furthermore, the operator L is self-adjoint. We invoke

$$\lim_{x \to \infty} u(x) = 0 \tag{3.108}$$

$$\lim_{x \to 0} u(x) \text{ finite} \tag{3.109}$$

We seek the spectral representation of the operator in (3.107) with the limiting conditions given in (3.108) and (3.109). The Green's function associated with this operator has been obtained previously in (2.184) and is repeated here for convenience, as follows:

$$g(x,\xi) = \frac{\pi}{2i}\begin{cases} H_0^{(2)}(\sqrt{\lambda}\xi)J_0(\sqrt{\lambda}x), & x < \xi \\ H_0^{(2)}(\sqrt{\lambda}x)J_0(\sqrt{\lambda}\xi), & x > \xi \end{cases} \tag{3.110}$$

where

$$\mathrm{Im}\sqrt{\lambda} < 0 \tag{3.111}$$

To assure the condition in (3.111), we restrict λ as follows:

$$\lambda = |\lambda|e^{i\phi}, \qquad -2\pi < \phi < 0 \tag{3.112}$$

so that

$$\sqrt{\lambda} = |\lambda|^{1/2}e^{i\phi/2}, \qquad -\pi < \frac{\phi}{2} < 0 \tag{3.113}$$

We may show that this definition produces a branch cut in $g(x, \xi, \lambda)$ along the positive-real axis in the λ-plane. Consider the case $x > \xi$. Approaching the positive-real axis from above, we have

$$\lim_{\phi\to-2\pi} H_0^{(2)}(\sqrt{\lambda}x)J_0(\sqrt{\lambda}\xi) = H_0^{(2)}(e^{-i\pi}|\lambda|^{1/2}x)J_0(e^{-i\pi}|\lambda|^{1/2}\xi) \tag{3.114}$$

But, using a well-known Bessel function identity [8], we have

$$J_0(e^{-i\pi}|\lambda|^{1/2}\xi) = J_0(|\lambda|^{1/2}\xi) \tag{3.115}$$

and using a well-known Hankel function identity [9], we have

$$H_0^{(2)}(e^{-i\pi}|\lambda|^{1/2}x) = -H_0^{(1)}(|\lambda|^{1/2}x) \tag{3.116}$$

so that

$$\lim_{\phi\to-2\pi} H_0^{(2)}(\sqrt{\lambda}x)J_0(\sqrt{\lambda}\xi) = -H_0^{(1)}(|\lambda|^{1/2}x)J_0(|\lambda|^{1/2}\xi) \tag{3.117}$$

On the other hand, approaching the positive-real axis from below, we have

$$\lim_{\phi\to 0} H_0^{(2)}(\sqrt{\lambda}x)J_0(\sqrt{\lambda}\xi) = H_0^{(2)}(|\lambda|^{1/2}x)J_0(|\lambda|^{1/2}\xi) \tag{3.118}$$

We note that (3.117) and (3.118) indicate a jump in the Green's function as we cross the positive real axis. To produce the spectral representation, we again consider the contour in Fig. 3-3 and write

$$\oint_{C_R+C_1+C_\rho+C_2} g(x,\xi,\lambda)d\lambda = 0 \tag{3.119}$$

In a manner similar to Example 3.3, we may show that the contribution along C_ρ vanishes as $\Gamma \to 0$ and $\rho \to 0$. We leave this for the reader to verify. Along $C_1 + C_2$, we have

$$\lim_{\rho\to 0}\lim_{R\to\infty}\lim_{\Gamma\to 0}\int_{C_1+C_2} g(x,\xi,\lambda)d\lambda = \frac{\pi}{2i}\left[\int_{\infty}^{0} H_0^{(2)}(|\lambda|^{1/2}x)J_0(|\lambda|^{1/2}\xi)d\lambda - \int_0^{\infty} H_0^{(1)}(|\lambda|^{1/2}x)J_0(|\lambda|^{1/2}\xi)d\lambda\right]$$
$$= -\frac{\pi}{i}\int_0^{\infty} J_0(|\lambda|^{1/2}x)J_0(|\lambda|^{1/2}\xi)d\lambda \tag{3.120}$$

Along C_R, we have

$$\lim_{\Gamma\to 0}\lim_{R\to\infty}\int_{C_R} g(x,\xi,\lambda)d\lambda = -2\pi i\frac{\delta(x-\xi)}{x} \tag{3.121}$$

Taking the appropriate limits in (3.119) and substituting (3.120) and (3.121), we obtain

$$\frac{\delta(x-\xi)}{x} = \frac{1}{2}\int_0^{\infty} J_0(|\lambda|^{1/2}x)J_0(|\lambda|^{1/2}\xi)d\lambda \tag{3.122}$$

Letting $k = |\lambda|^{1/2}$, we find that

$$\frac{\delta(x-\xi)}{x} = \int_0^{\infty} J_0(kx)J_0(k\xi)kdk \tag{3.123}$$

which is the required spectral representation. Although this representation has been obtained with the restriction $x > \xi$, the restriction can now be removed in the same manner as in Example 3.3 because of the symmetry of the Green's function.

The representation in (3.123) leads to the *Fourier–Bessel Transform* of order zero. Indeed, consider a Hilbert space $\mathcal{L}_2(0,\infty)$ with inner product

$$\langle s, t\rangle = \int_0^{\infty} s(x)t(x)xdx \tag{3.124}$$

For any $s(x) \in \mathcal{L}_2(0,\infty)$, we have

$$s(x) = \int_0^{\infty} s(\xi)\frac{\delta(x-\xi)}{x}\xi d\xi \tag{3.125}$$

Substituting (3.123) into (3.125), we produce the Fourier–Bessel transform pair

$$S(k) = \int_0^{\infty} s(x)J_0(kx)xdx \tag{3.126}$$

$$s(x) = \int_0^\infty S(k) J_0(kx) k dk \tag{3.127}$$

Symbolically, we write

$$s(x) \Longleftrightarrow S(k)$$

A very useful relation is obtained by noting that for $u(x) \in \mathcal{L}_2(0, \infty)$,

$$\begin{aligned}\frac{1}{x}\left[\frac{d}{dx}\left(x\frac{du}{dx}\right)\right] &= \int_0^\infty U(k)\left\{\frac{1}{x}\frac{d}{dx}\left[x\frac{dJ_0(kx)}{dx}\right]\right\} kdk \\ &= \int_0^\infty [-k^2 U(k)] J_0(kx) k dk \end{aligned} \tag{3.128}$$

Therefore,

$$\frac{1}{x}\left[\frac{d}{dx}\left(x\frac{du}{dx}\right)\right] \Longleftrightarrow -k^2 U(k) \tag{3.129}$$

We note that the result in (3.128) depends on the interchange of differentiation and integration. This operation can be justified by using the procedure followed in (3.76)–(3.79) and in (3.102)–(3.104). The details are left for the problems.

■

EXAMPLE 3.6 We wish to find the spectral representation of the operator

$$L = -x\left[\frac{d}{dx}\left(x\frac{d}{dx}\right)\right] - (kx)^2, \qquad \text{Im}(k) < 0 \tag{3.130}$$

on $\mathcal{L}_2(0, \infty)$. By examining (2.22), we identify $p(x) = x$ and $w(x) = 1/x$. The Green's function differential equation associated with L_λ is given by

$$-x\left[\frac{d}{dx}\left(x\frac{dg}{dx}\right)\right] - (kx)^2 g - \lambda g = x\delta(x - \xi) \tag{3.131}$$

where we have identified $\delta(x - \xi)/w(x) = x\delta(x - \xi)$. We investigate limit point and limit circle conditions as $x \to 0$ and as $x \to \infty$ by examining solutions to the homogeneous equation

$$-x\left[\frac{d}{dx}\left(x\frac{du}{dx}\right)\right] - (kx)^2 u - \lambda u = 0$$

For $\lambda = 0$, two independent solutions are given by

$$u_1 = H_0^{(2)}(kx)$$

and

$$u_2 = H_0^{(1)}(kx)$$

Let ξ be an arbitrary interior point on the interval $x \in (0, \infty)$. As $x \to 0$, both u_1 and u_2 are logarithmically singular. The singularity is weak enough, however, that they are both in $\mathcal{L}_2(0, \xi)$. We therefore have the limit circle case as $x \to 0$. The solution u_2 diverges exponentially as $x \to \infty$, and thus is not in $\mathcal{L}_2(\xi, \infty)$. We therefore have the limit point case as $x \to \infty$. We invoke the limit conditions

$$\lim_{x\to\infty} u(x) = 0 \tag{3.132}$$

$$\lim_{x\to 0} u(x) \text{ finite} \tag{3.133}$$

The limit conditions associated with the Green's function are

$$\lim_{x\to\infty} g(x, \xi) = 0 \tag{3.134}$$

$$\lim_{x\to 0} g(x, \xi) \text{ finite} \tag{3.135}$$

Define a parameter ν by

$$\nu = e^{i\pi/2}\sqrt{\lambda} = i\sqrt{\lambda} \tag{3.136}$$

Then, for $x \neq \xi$, we have

$$x\left[\frac{d}{dx}\left(x\frac{dg}{dx}\right)\right] + \left[(kx)^2 - \nu^2\right]g = 0 \tag{3.137}$$

We identify (3.137) as Bessel's equation of order ν and argument kx [10]. The Green's function can be composed of linear combinations of various Bessel functions as follows:

$$g = \begin{cases} AJ_\nu(kx) + CJ_{-\nu}(kx), & x < \xi \\ BH_\nu^{(2)}(kx) + DH_\nu^{(1)}(kx), & x > \xi \end{cases} \tag{3.138}$$

where J_ν, $J_{-\nu}$, $H_\nu^{(2)}$, and $H_\nu^{(1)}$ are linearly independent solutions to Bessel's equation of order ν [10]. For $\text{Im}(k) < 0$, $H_\nu^{(1)}$ diverges as $x \to \infty$ [11]. Therefore, $D = 0$. We may set $C = 0$ by using the following argument. For $x \to 0$, we have

$$J_\nu(kx) \longrightarrow \frac{(kx)^\nu}{2^\nu \nu!}$$

$$J_{-\nu}(kx) \longrightarrow \frac{2^\nu (kx)^{-\nu}}{(-\nu)!}$$

Therefore, if we choose $\text{Re}(\nu) > 0$, $J_{-\nu}$ diverges as $x \to 0$ and we must set $C = 0$. The Green's function can now be written as follows:

$$g = \begin{cases} AJ_\nu(kx), & x < \xi \\ BH_\nu^{(2)}(kx), & x > \xi \end{cases} \tag{3.139}$$

The evaluation of the coefficients A and B proceeds in a manner similar to that in Example 2.21. Invoking the continuity and jump conditions at $x = \xi$, we find that

$$A = \frac{\pi}{2i} H_\nu^{(2)}(k\xi)$$

$$B = \frac{\pi}{2i} J_\nu(k\xi)$$

Substitution into (3.139) gives

$$g = \frac{\pi}{2i} \begin{cases} H_\nu^{(2)}(k\xi) J_\nu(kx), & x < \xi \\ H_\nu^{(2)}(kx) J_\nu(k\xi), & x > \xi \end{cases} \tag{3.140}$$

where

$$\text{Im}(k) < 0 \tag{3.141}$$

$$\text{Re}(\nu) > 0 \tag{3.142}$$

The last step in the determination of the Green's function involves making the transformation from ν to λ in accordance with (3.136). If we define

$$\lambda = |\lambda| e^{i\phi}, \qquad 0 > \phi > -2\pi \tag{3.143}$$

then

$$\sqrt{\lambda} = |\lambda|^{1/2} e^{i\phi/2}, \qquad 0 > \frac{\phi}{2} > -\pi \tag{3.144}$$

This result implies that

$$\text{Im}\sqrt{\lambda} < 0 \tag{3.145}$$

Substituting (3.144) into (3.136) gives

$$\nu = |\lambda|^{1/2} e^{i(\phi+\pi)/2}, \qquad \frac{\pi}{2} > \frac{\phi + \pi}{2} > -\frac{\pi}{2} \tag{3.146}$$

The angular range in (3.146) is consistent with the restriction on ν in (3.142). We therefore have

$$g(x, \xi, \lambda) = \frac{\pi}{2i} \begin{cases} H_{i\sqrt{\lambda}}^{(2)}(k\xi) J_{i\sqrt{\lambda}}(kx), & x < \xi \\ H_{i\sqrt{\lambda}}^{(2)}(kx) J_{i\sqrt{\lambda}}(k\xi), & x > \xi \end{cases} \tag{3.147}$$

where the branch cut in $\sqrt{\lambda}$ lies along the positive-real axis and is explicitly determined by (3.143).

Our next step is to determine the spectral representation of $x\delta(x - \xi)$ by integrating over the Green's function with respect to λ in a similar manner to that performed in Example 3.5. We first consider the case $x < \xi$. We find that the branch cut in $\sqrt{\lambda}$, defined in (3.143), produces a branch cut in $g(x, \xi, \lambda)$ along the positive-real axis. Indeed,

$$\lim_{\phi \to -2\pi} H^{(2)}_{i\sqrt{\lambda}}(k\xi) J_{i\sqrt{\lambda}}(kx)\, d = H^{(2)}_{-i|\lambda|^{1/2}}(k\xi) J_{-i|\lambda|^{1/2}}(kx)$$

$$\lim_{\phi \to 0} H^{(2)}_{i\sqrt{\lambda}}(k\xi) J_{i\sqrt{\lambda}}(kx) = H^{(2)}_{i|\lambda|^{1/2}}(k\xi) J_{i|\lambda|^{1/2}}(kx)$$

Since J_ν and $J_{-\nu}$ are linearly independent for any $\nu \in \mathbf{C}$, we conclude that there is a jump in the Green's function across the positive-real axis, resulting in a branch cut. The appropriate contour is the one shown in Fig. 3-4. Substituting (3.147) into (3.82), we find for $x < \xi$

$$2\pi i x \delta(x - \xi)$$

$$= \frac{\pi}{2i} \left[\int_\infty^0 H^{(2)}_{i|\lambda|^{1/2}}(k\xi) J_{i|\lambda|^{1/2}}(kx) d\lambda + \int_0^\infty H^{(2)}_{-i|\lambda|^{1/2}}(k\xi) J_{-i|\lambda|^{1/2}}(kx) d\lambda \right] \tag{3.148}$$

But [12],

$$H^{(2)}_{-i|\lambda|^{1/2}} = e^{-i\pi(i|\lambda|^{1/2})} H^{(2)}_{i|\lambda|^{1/2}} \tag{3.149}$$

Substituting (3.149) into (3.148) and combining integrals gives

$$-4x\delta(x - \xi) = \textstyle\int_0^\infty e^{-i\pi(i|\lambda|^{1/2})} H^{(2)}_{i|\lambda|^{1/2}}(k\xi)$$

$$\cdot \left[J_{-i|\lambda|^{1/2}}(kx) - e^{i\pi(i|\lambda|^{1/2})} J_{i|\lambda|^{1/2}}(kx) \right] d\lambda \tag{3.150}$$

But for any $\nu \in \mathbf{C}$ [13],

$$J_{-\nu}(z) - e^{i\pi\nu} J_\nu(z) = \frac{H^{(2)}_\nu(z)}{i \csc(\nu\pi)} \tag{3.151}$$

Substitution into (3.150) gives

$$4x\delta(x - \xi) = i \int_0^\infty e^{-i\pi(i|\lambda|^{1/2})} \sin(i\pi|\lambda|^{1/2}) H^{(2)}_{i|\lambda|^{1/2}}(k\xi) H^{(2)}_{i|\lambda|^{1/2}}(kx) d\lambda \tag{3.152}$$

Let

$$\beta = i|\lambda|^{1/2} \tag{3.153}$$

Then,

$$x\delta(x - \xi) = \frac{1}{4} \int_0^{i\infty} \left(e^{-i2\pi\beta} - 1 \right) H^{(2)}_\beta(kx) H^{(2)}_\beta(k\xi) \beta d\beta \tag{3.154}$$

We note that this result is not altered by interchanging x and ξ. Therefore, our restriction $x < \xi$ can be removed. Equation (3.154) gives the spectral representation of the delta function for the operator defined in (3.130)–(3.133).

The representation in (3.154) leads to the *Kantorovich–Lebedev Transform* [14],[15]. Indeed, consider a Hilbert space $\mathcal{L}_2(0, \infty)$ with inner product

$$\langle s, t\rangle = \int_0^\infty s(x)t(x)\frac{dx}{x} \tag{3.155}$$

For any $s(x) \in \mathcal{L}_2(0, \infty)$, we have

$$s(x) = \int_0^\infty s(\xi)x\delta(x-\xi)\frac{d\xi}{\xi} \tag{3.156}$$

Substituting (3.154) into (3.156), we produce the Kantorovich–Lebedev transform pair

$$F(\beta) = \int_0^\infty f(x)H_\beta^{(2)}(kx)\frac{dx}{x} \tag{3.157}$$

$$f(x) = \frac{1}{4}\int_0^{i\infty} \left(e^{-i2\pi\beta} - 1\right) F(\beta)H_\beta^{(2)}(kx)\beta d\beta \tag{3.158}$$

Alternately, we can manipulate (3.158) to produce

$$f(x) = \frac{1}{4}\int_{i\infty}^{-i\infty} F(\beta)H_\beta^{(2)}(kx)\beta d\beta \tag{3.159}$$

The details of producing (3.159) from (3.158) are left for Problem 3.6. We indicate the Kantorovich–Lebedev transform relationship by

$$f(x) \Longleftrightarrow F(\beta) \tag{3.160}$$

If we apply the operator L in (3.130) to both sides of (3.159), we produce the useful relationship

$$\left\{-x\left[\frac{d}{dx}\left(x\frac{d}{dx}\right)\right] - (kx)^2\right\} f(x) \Longleftrightarrow -\beta^2 F(\beta) \tag{3.161}$$

The interchange of differentiation and integration used to produce (3.161) can be justified in the same manner as in the procedure in (3.76)–(3.79) and in (3.102)–(3.104). The details are left for the problems. The Kantorovich–Lebedev transform is useful in solving certain electromagnetic problems in cylindrical coordinates, as we shall discover in the next chapter.

■

In the mathematical literature, the spectral contribution resulting from pole contributions, such as in (3.47), is called the *discrete spectrum*, whereas the contribution from the branch cut, such as in (3.64), is called the *continuous spectrum*. We next inquire if it is possible to have both a continuous and discrete spectrum associated with an operator. The example we choose involves an operator that is not self-adjoint. The theory of nonself-adjoint operators is both difficult and incomplete. However, in the simple example to follow, we are able to obtain the spectral representation in a straightforward manner.

EXAMPLE 3.7 We consider the spectral representation of the operator $L = -d^2/dx^2$, with boundary and limiting conditions

$$\lim_{x\to\infty} u(x) = 0 \tag{3.162}$$

$$u'(0) = \alpha u(0), \qquad \text{Re}(\alpha) < 0 \tag{3.163}$$

Since α is complex, the operator L is nonself-adjoint. The associated Green's function problem is given by

$$-\frac{d^2 g}{dx^2} - \lambda g = \delta(x - \xi)$$

$$\frac{dg(0,\xi)}{dx} = \alpha g(0,\xi)$$

$$\lim_{x\to\infty} g(x,\xi) = 0$$

We have previously obtained this Green's function in Example 2.23. We repeat the result given in (2.216) for convenience, viz.

$$g(x,\xi) = \frac{1}{i\sqrt{\lambda}+\alpha}\begin{cases} e^{-i\sqrt{\lambda}x}(\cos\sqrt{\lambda}\xi + \dfrac{\alpha}{\sqrt{\lambda}}\sin\sqrt{\lambda}\xi), & x > \xi \\ e^{-i\sqrt{\lambda}\xi}(\cos\sqrt{\lambda}x + \dfrac{\alpha}{\sqrt{\lambda}}\sin\sqrt{\lambda}x), & x < \xi \end{cases} \tag{3.164}$$

where

$$\text{Im}\sqrt{\lambda} < 0 \tag{3.165}$$

The restriction in (3.165) can again be assured by defining $\sqrt{\lambda}$ as in (3.56) and (3.57) so that, once again (Fig. 3-2),

$$\lim_{\phi\to 2\pi} \sqrt{\lambda} = -|\lambda|^{1/2}$$

$$\lim_{\phi\to 4\pi} \sqrt{\lambda} = |\lambda|^{1/2}$$

The branch cut in $\sqrt{\lambda}$ along the positive-real axis results in a branch cut in the same location in $g(x,\xi,\lambda)$. Indeed, for $x < \xi$, we obtain in (3.164)

$$\lim_{\phi\to 2\pi} g(x,\xi,\lambda) = \frac{e^{i|\lambda|^{1/2}\xi}}{\alpha - i|\lambda|^{1/2}}(\cos|\lambda|^{1/2}x + \frac{\alpha}{|\lambda|^{1/2}}\sin|\lambda|^{1/2}x)$$

$$\lim_{\phi\to 4\pi} g(x,\xi,\lambda) = \frac{e^{-i|\lambda|^{1/2}\xi}}{\alpha + i|\lambda|^{1/2}}(\cos|\lambda|^{1/2}x + \frac{\alpha}{|\lambda|^{1/2}}\sin|\lambda|^{1/2}x)$$

The result for $x > \xi$ is obtained by interchanging x and ξ. In addition to the branch cut on the positive-real axis, $g(x, \xi, \lambda)$ has an isolated singularity at the location λ_0, given by solving

$$i\sqrt{\lambda} + \alpha = 0$$

with the result

$$\lambda_0 = -\alpha^2$$

We now show that this singularity is on the proper Riemann sheet. Indeed, we have $\sqrt{\lambda_0} = i\alpha$. From this relationship, we easily find that the relation $\mathrm{Im}\sqrt{\lambda} < 0$ implies that $\mathrm{Re}(\alpha) < 0$, as assumed in the problem statement.

We now show that this singularity is a simple pole. For $x < \xi$, we write

$$g(x, \xi, \lambda) = f_1(\lambda) f_2(x, \xi, \lambda) \tag{3.166}$$

where

$$f_1(\lambda) = \frac{1}{i\sqrt{\lambda} + \alpha} \tag{3.167}$$

$$f_2(x, \xi, \lambda) = e^{-i\sqrt{\lambda}\xi}(\cos\sqrt{\lambda}x + \frac{\alpha}{\sqrt{\lambda}} \sin\sqrt{\lambda}x) \tag{3.168}$$

We note that $f_2(x, \xi, \lambda)$ is regular at λ_0. Consider

$$f_1(\lambda) = \frac{p(\lambda)}{q(\lambda)}$$

where $p(\lambda) = 1$ and

$$q(\lambda) = i\sqrt{\lambda} + \alpha$$

We have

$$q(-\alpha^2) = 0$$

$$q'(-\alpha^2) = \frac{1}{2\alpha}$$

We conclude that $f_1(\lambda)$ has a simple pole at λ_0 with residue

$$\mathrm{Res}\{f_1(\lambda); \lambda_0\} = \frac{p(\lambda_0)}{q'(\lambda_0)} = 2\alpha$$

To obtain the spectral representation, we integrate the Green's function $g(x, \xi, \lambda)$ around the closed contour shown in Fig. 3-5. For $x > \xi$, we have

$$\begin{aligned}\oint g(x, \xi, \lambda) d\lambda &= 2\pi i \mathrm{Res}\{g(x, \xi, \lambda); \lambda_0\} \\ &= 2\pi i e^{\alpha x}[\cos(i\alpha\xi) - i\sin(i\alpha\xi)]\mathrm{Res}\{f_1(\lambda); \lambda_0\} \\ &= 2\pi i(2\alpha)e^{\alpha(x+\xi)} \end{aligned} \tag{3.169}$$

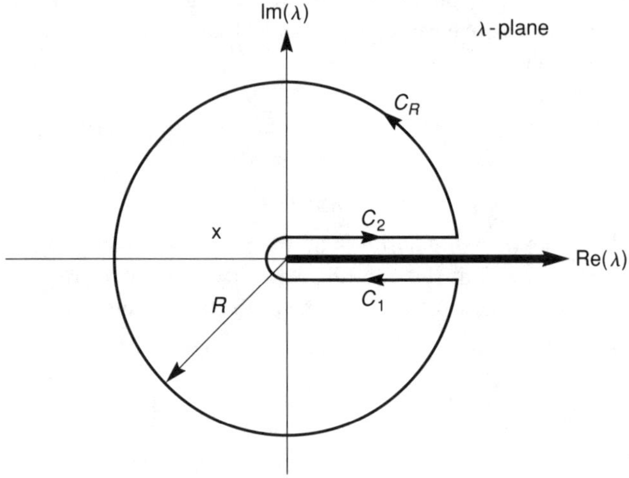

Fig. 3-5 Contour for evaluation of the spectral representation for Example 3.4. Contour includes the simple pole (×) at λ_0.

The integral around the closed contour (Fig. 3-5) consists of the integral around the circle of radius R plus the integrals along either side of the branch cut. Therefore, in the limit as $R \to \infty$, we obtain

$$-\delta(x-\xi) + \frac{1}{2\pi i} \lim_{R\to\infty} \int_{C_1+C_2} g(x,\xi,\lambda)d\lambda = 2\alpha e^{\alpha(x+\xi)} \tag{3.170}$$

Evaluating the integrals along C_1 and C_2 in a similar manner to the process in Example 3.3, we obtain

$$\lim_{R\to\infty} \int_{C_1+C_2} = 2i \int_0^\infty (\cos|\lambda|^{1/2}x + \frac{\alpha}{|\lambda|^{1/2}} \sin|\lambda|^{1/2}x)$$
$$\cdot (\cos|\lambda|^{1/2}\xi + \frac{\alpha}{|\lambda|^{1/2}} \sin|\lambda|^{1/2}\xi) \frac{|\lambda|^{1/2}d\lambda}{\alpha^2+|\lambda|} \tag{3.171}$$

We let $k = |\lambda|^{1/2}$ and obtain for the spectral representation

$$\delta(x-\xi) = -2\alpha e^{\alpha(x+\xi)} + \frac{2}{\pi}\int_0^\infty (\cos kx + \frac{\alpha}{k}\sin kx)(\cos k\xi + \frac{\alpha}{k}\sin k\xi)\frac{k^2dk}{\alpha^2+k^2} \tag{3.172}$$

The first term on the right side gives the discrete spectral contribution, while the second term gives the continuous spectral contribution. Although the delta function representation has been obtained with the restriction $x > \xi$, we note in (3.164) that the case $x < \xi$ can be obtained by interchanging x and ξ. Since such an interchange leaves the result in (3.172) unaltered, the restriction can be removed.

In Problem 5.6 given at the end of Chapter 5, the spectral representation in (3.172) will be used to characterize a source over a flat surface characterized by a surface impedance. We shall discover that we may associate the first term with a surface wave bound to the surface and the second term with radiation carried away from the surface.

We remark that we assumed $\mathrm{Re}(\alpha) < 0$ in the problem statement. In addition, we found that $\mathrm{Im}\sqrt{\lambda} < 0$ implies $\mathrm{Re}(\alpha) < 0$, as required. If the problem had stated that $\mathrm{Re}(\alpha) \geq 0$, the pole at $\lambda_0 = -\alpha^2$ would have been on the improper Riemann sheet and the discrete spectral term in (3.172) would be missing.

In a similar manner to that in Examples 3.2 and 3.3, we now cast (3.172) in a form that we shall call the *impedance transform*. For $u(x) \in \mathcal{L}_2(0, \infty)$, we have

$$u(x) = \int_0^\infty \delta(x - \xi) u(\xi) d\xi \tag{3.173}$$

Substituting (3.172) and interchanging the order of integration, we obtain

$$u(x) = -2\alpha e^{\alpha x} U_0 + \frac{2}{\pi} \int_0^\infty U(k)(\cos kx + \frac{\alpha}{k} \sin kx) \frac{k^2 dk}{k^2 + \alpha^2} \tag{3.174}$$

where

$$U_0 = \int_0^\infty u(x) e^{\alpha x} dx \tag{3.175}$$

$$U(k) = \int_0^\infty u(x)(\cos kx + \frac{\alpha}{k} \sin kx) dx \tag{3.176}$$

Equations (3.175) and (3.176) comprise the impedance transform, yielding the spectral coefficients U_0 and $U(k)$. Equation (3.174) is the inverse impedance transform. We call (3.175) the *zeroth-order impedance transform*, while (3.176) is the *kth-order impedance transform*. In (3.174), we identify

$$e^{\alpha x}$$

as an eigenfunction of the operator $-d^2/dx^2$ with boundary conditions given in (3.162) and (3.163). In addition,

$$\cos kx + \frac{\alpha}{k} \sin kx$$

is an improper eigenfunction of the same operator with the same boundary conditions. We define an inner product for the space by

$$\langle u, v \rangle = \int_0^\infty u(x) \overline{v}(x) dx$$

With this definition, we may write (3.175) and (3.176) as follows:

$$U_0 = \langle u, e^{\overline{\alpha} x} \rangle \tag{3.177}$$

$$U(k) = \langle u, \cos kx + \frac{\overline{\alpha}}{k} \sin kx \rangle \tag{3.178}$$

We identify

$$e^{\overline{\alpha}x}$$

as an adjoint eigenfunction of the operator $-d^2/dx^2$ with adjoint boundary conditions given by

$$\begin{aligned} v'(0) &= \overline{\alpha} v(0) \\ \lim_{x \to \infty} v(x) &= 0 \end{aligned} \tag{3.179}$$

In addition,

$$\cos kx + \frac{\overline{\alpha}}{k} \sin kx$$

is an improper adjoint eigenfunction of the same operator with the adjoint boundary condition given in (3.179).

We next use the impedance transform to solve the following SLP3 problem:

$$-u'' - \lambda u = f, \qquad \lambda \in \mathbf{C} \tag{3.180}$$

with boundary condition

$$u'(0) = \alpha u(0), \qquad \mathrm{Re}(\alpha) < 0 \tag{3.181}$$

This problem is in the limit point case as $x \to \infty$. We therefore invoke the limiting condition

$$\lim_{x \to \infty} u(x) = 0 \tag{3.182}$$

In order to solve the differential equation in (3.180) by the use of the impedance transform, we require the zeroth-order and kth-order impedance transform of $-d^2u/dx^2$. For the zeroth order, we have

$$\langle -u'', e^{\overline{\alpha}x} \rangle = \langle u, -\frac{d^2}{dx^2} e^{\overline{\alpha}x} \rangle + \left(-u' e^{\alpha x} + u \frac{d}{dx} e^{\alpha x} \rangle \right) \Bigg|_0^\infty = -\alpha^2 \langle u, e^{\overline{\alpha}x} \rangle = -\alpha^2 U_0 \tag{3.183}$$

where we have used (3.181), (3.182), and

$$\lim_{x \to \infty} e^{\alpha x} = 0$$

Symbolically, we indicate this zeroth-order transform by

$$-u'' \overset{0}{\Longrightarrow} -\alpha^2 U_0 \tag{3.184}$$

Similarly, for the kth-order transform, we have

$$
\begin{aligned}
\langle -u'', \cos kx + \frac{\overline{\alpha}}{k} \sin kx \rangle &= \langle u, -\frac{d^2}{dx^2}(\cos kx + \frac{\overline{\alpha}}{k} \sin kx)\rangle \\
&\quad + \left[-u'(\cos kx + \frac{\alpha}{k} \sin kx) + u\frac{d}{dx}(\cos kx + \frac{\alpha}{k} \sin kx) \right] \Bigg|_0^\infty \\
&= k^2 \langle u, \cos kx + \frac{\overline{\alpha}}{k} \sin kx \rangle = k^2 U(k)
\end{aligned}
\tag{3.185}
$$

where we have used (3.181) and (3.182). In addition, we have used the fact that, in the usual cases in electromagnetics,

$$\lim_{x\to\infty} u(x) = 0$$

implies that

$$\lim_{x\to\infty} u'(x) = 0$$

Symbolically, we indicate this transform by

$$-u'' \stackrel{k}{\Longrightarrow} k^2 U(k) \tag{3.186}$$

We now apply the zeroth-order transform to (3.180) to give

$$-(\alpha^2 + \lambda)U_0 = F_0$$

where F_0 is the zeroth-order transform of $f(x)$. Rearranging, we have

$$U_0 = -\frac{F_0}{\alpha^2 + \lambda} \tag{3.187}$$

Similarly, for the kth-order transform, we find that

$$(k^2 - \lambda)U(k) = F(k)$$

where $F(k)$ is the kth-order transform of $f(x)$. Rearranging, we have

$$U(k) = \frac{F(k)}{k^2 - \lambda} \tag{3.188}$$

Substituting these results into (3.174) gives the final result, viz.

$$u(x) = \frac{2\alpha e^{\alpha x} F_0}{\alpha^2 + \lambda} + \frac{2}{\pi} \int_0^\infty F(k) \frac{(\cos kx + \frac{\alpha}{k} \sin kx)}{k^2 - \lambda} \frac{k^2 dk}{k^2 + \alpha^2} \tag{3.189}$$

■

We have now concluded our presentation of the spectral representation method. This method and the Green's function method together comprise a powerful tool for the solution of many of the partial differential equations found in electromagnetic radiation, scattering, and diffraction. In subsequent chapters, we shall study electromagnetic source representations, and then develop the solution methods for a large class of electromagnetic problems. We shall conclude this chapter with a brief discussion of the connection between the Green's function method and the spectral representation method.

3.5 GREEN'S FUNCTIONS AND SPECTRAL REPRESENTATIONS

There is an important connection between the Green's function method and the spectral representation method. Indeed, consider the result in Example 3.3, given in (3.81), viz.

$$u(x) = \frac{2}{\pi}\int_0^\infty \frac{F(k)}{k^2-\lambda}\sin kx dk \tag{3.190}$$

where

$$F(k) = \int_0^\infty f(\xi)\sin k\xi d\xi \tag{3.191}$$

If we substitute (3.191) into (3.190) and interchange the order of integration, we obtain

$$u(x) = \int_0^\infty f(\xi)\left[\frac{2}{\pi}\int_0^\infty \frac{\sin kx \sin k\xi}{k^2-\lambda}dk\right]d\xi \tag{3.192}$$

We identify the term in square brackets as the Green's function

$$g(x,\xi) = \frac{2}{\pi}\int_0^\infty \frac{\sin kx \sin k\xi}{k^2-\lambda}dk \tag{3.193}$$

We compare this result with the Green's function obtained in Example 2.18, given in (2.172), viz.

$$g(x,\xi) = \frac{1}{\sqrt{\lambda}}\begin{cases} e^{-i\sqrt{\lambda}\xi}\sin\sqrt{\lambda}x, & x<\xi \\ e^{-i\sqrt{\lambda}x}\sin\sqrt{\lambda}\xi, & x>\xi \end{cases} \tag{3.194}$$

Although (3.193) and (3.194) appear very different, they are different representations of the same Green's function. Indeed, comparing (3.193) and (3.194), we must have

$$\int_0^\infty \frac{\sin kx \sin k\xi}{k^2-\lambda}dk = \frac{\pi}{2\sqrt{\lambda}}\begin{cases} e^{-i\sqrt{\lambda}\xi}\sin\sqrt{\lambda}x, & x<\xi \\ e^{-i\sqrt{\lambda}x}\sin\sqrt{\lambda}\xi, & x>\xi \end{cases} \tag{3.195}$$

as we could easily verify by the calculus of residues. Since (3.193) requires an integration and (3.194) does not, it would appear that the spectral representation is not as useful in practice. Its utility, however, becomes clear as soon as we begin considering partial rather than ordinary differential equations in subsequent chapters.

PROBLEMS

3.1. For the operator $L = -d^2/dx^2$ and boundary conditions $u(0) = u(a) = 0$, begin with the Green's function for L_λ and show that the spectral representation of the delta function is given by

$$\delta(x-\xi) = \frac{2}{a}\sum_{n=1}^{\infty} \sin\frac{n\pi x}{a} \sin\frac{n\pi\xi}{a}$$

Use this spectral representation to obtain the solution to the differential equation

$$L_\lambda u = f$$

with the operator L and the boundary conditions given above.

3.2. For the operator $L = -d^2/dx^2$ and boundary conditions $u(0) = u(2\pi)$ and $u'(0) = u'(2\pi)$, the Green's function for L_λ was found in Problem 2.18. The result is repeated here for reference, viz.

$$g(x,\xi) = -\frac{1}{2\sqrt{\lambda}\sin\sqrt{\lambda}\pi}\begin{cases} \cos\sqrt{\lambda}(\xi - x - \pi), & x < \xi \\ \cos\sqrt{\lambda}(x - \xi - \pi), & x > \xi \end{cases}$$

Beginning with this Green's function, show that the spectral representation of the delta function is given by

$$\delta(x-\xi) = \frac{1}{2\pi} + \frac{1}{\pi}\sum_{n=1}^{\infty}(\cos nx \cos n\xi + \sin nx \sin n\xi)$$

By using Euler's identity, show that an alternate representation is given by

$$\delta(x-\xi) = \frac{1}{2\pi}\sum_{n=-\infty}^{\infty} e^{in(x-\xi)}$$

Show that this alternate representation leads to the complex Fourier series

$$f(x) = \sum_{n=-\infty}^{\infty} a_n \sqrt{\frac{1}{2\pi}}e^{inx}$$

$$a_n = \int_0^{2\pi} f(x)\sqrt{\frac{1}{2\pi}}e^{-inx}\,dx$$

3.3. For the operator $L = -d^2/dx^2$ with boundary condition $u'(0) = 0$ and limiting condition

$$\lim_{x \to \infty} u(x) = 0$$

begin with the Green's function for L_λ and show that the spectral representation of the delta function is given by

$$\delta(x - \xi) = \frac{2}{\pi} \int_0^\infty \cos kx \cos k\xi \, dk$$

Use this spectral representation to obtain the solution to the differential equation

$$L_\lambda u = f$$

with the operator L and the boundary conditions given above.

3.4. In Example 3.5, it was shown that

$$\frac{1}{x}\left[\frac{d}{dx}\left(x\frac{du}{dx}\right)\right] = \int_0^\infty U(k)\left\{\frac{1}{x}\frac{d}{dx}\left[x\frac{dJ_0(kx)}{dx}\right]\right\} k\,dk$$

$$= \int_0^\infty [-k^2 U(k)] J_0(kx) k\,dk$$

and therefore,

$$\frac{1}{x}\left[\frac{d}{dx}\left(x\frac{du}{dx}\right)\right] \Longleftrightarrow -k^2 U(k)$$

This result involves the interchange of differential operator and integration. Justify this result by following the procedure used in (3.76)–(3.79) and in (3.102)–(3.104).

3.5. Consider the following Green's function problem associated with Bessel's equation of order ν:

$$(L - \lambda) g = \frac{\delta(x - \xi)}{x}$$

where

$$L = -\frac{1}{x}\left[\frac{d}{dx}\left(x\frac{d}{dx}\right)\right] + \frac{\nu^2}{x^2}$$

with limiting condition

$$\lim_{x \to \infty} g(x, \xi) = 0$$

where λ is complex. By invoking the condition that g must be finite as $x \to 0$, show that

$$g(x, \xi, \lambda) = \frac{\pi}{2i} \begin{cases} H_\nu^{(2)}(\sqrt{\lambda}\xi) J_\nu(\sqrt{\lambda}x), & x < \xi) \\ H_\nu^{(2)}(\sqrt{\lambda}x) J_\nu(\sqrt{\lambda}\xi), & x > \xi) \end{cases}$$

Note: This result can be obtained directly from Example 3.6 by identifying λ above with k^2 in (3.131) in Example 3.6. By integrating the Green's function around the closed contour in the complex λ-plane, as described in (3.119), show that the spectral representation of the delta function for the operator L and boundary conditions given above is given by

$$\frac{\delta(x-\xi)}{x} = \int_0^\infty J_\nu(kx)J_\nu(k\xi)kdk$$

Show that this representation leads to the *Fourier–Bessel Transform* of order ν, given by the pair

$$S(k) = \int_0^\infty s(x)J_\nu(kx)xdx$$

$$s(x) = \int_0^\infty S(k)J_\nu(kx)kdk$$

Produce the following Fourier–Bessel transform pair:

$$\left\{-\frac{1}{x}\left[\frac{d}{dx}\left(x\frac{d}{dx}\right)\right] + \frac{\nu^2}{x^2}\right\} s(x) \Longleftrightarrow k^2 S(k)$$

3.6. In the development of the Kantorovich–Lebedev transform, carefully complete the steps necessary to produce (3.159) from (3.158).

3.7. Justify the result in (3.161) by following the procedure used in (3.76)–(3.79) and (3.102)–(3.104).

3.8. In the development of the Kantorovich–Lebedev transform in Example 3.6, we considered the operator

$$L = -x\left[\frac{d}{dx}\left(x\frac{d}{dx}\right)\right] - (kx)^2, \qquad \text{Im}(k) < 0$$

where

$$\lim_{x\to\infty} u(x) = 0$$

$$\lim_{x\to 0} u(x) \text{ finite}$$

In [14], Stakgold considers the Kantorovich–Lebedev problem for a slightly different operator, viz.

$$L_s = -x\left[\frac{d}{dx}\left(x\frac{d}{dx}\right)\right] + \mu x^2, \qquad \mu > 0$$

His resulting spectral representation is given by

$$x\delta(x-\xi) = \frac{2}{\pi^2}\int_0^\infty \sinh \pi\beta K_{i\beta}(\sqrt{\mu}x)K_{i\beta}(\sqrt{\mu}\xi)\beta d\beta \qquad (3.196)$$

where $K_\lambda(y)$ is the modified Bessel function of the second kind, λth order [16]. By making the appropriate transformation between k and μ, show that the spectral representation in (3.154) transforms to (3.196).

REFERENCES

[1] Stakgold, I. (1979), *Green's Functions and Boundary Value Problems.* New York: Wiley Interscience, 414–415.

[2] Carrier, G.F., M. Krook, and C.E. Pearson (1966), *Functions of a Complex Variable.* New York: McGraw-Hill, 56–59.

[3] Friedman, B. (1956), *Principles and Techniques of Applied Mathematics.* New York: Wiley, 213–222.

[4] Churchill, R.V. (1960), *Complex Variables and Applications.* New York: McGraw-Hill, 159–161.

[5] Arfken, G. (1985), *Mathematical Methods for Physicists*, 3rd Edition. New York: Academic, 529–530.

[6] op.cit. Friedman, 234–237.

[7] Sneddon, I.N. (1972), *The Use of Integral Transforms.* New York: McGraw-Hill, 46–49.

[8] Abramowitz, M., and I.A. Stegun (Eds.) (1964), *Handbook of Mathematical Functions*, National Bureau of Standard, Applied Mathematics Series, 55, Superintendent of Documents, U.S. Government Printing Office, Washington, DC 20402, 361, eq. 9.1.35.

[9] Ibid., eq. 9.1.39.

[10] Ibid., 358, sect. 9.1.1.

[11] Harrington, R.F. (1961, reissued 1987), *Time-Harmonic Electromagnetic Fields.* New York: McGraw-Hill, 461–462.

[12] op.cit. Abramowitz and Stegun, eq. 9.1.6.

[13] Ibid., eq. 9.1.4.

[14] Stakgold, I. (1967), *Boundary Value Problems of Mathematical Physics.* New York: Macmillan, 317.

[15] Felsen, L.B., and N. Marcuvitz (1973), *Radiation and Scattering of Waves.* Englewood-Cliffs, NJ: Prentice-Hall, 323–326.

[16] op.cit. Harrington, 363.

4

Electromagnetic Sources

4.1 INTRODUCTION

In this chapter, we introduce electromagnetic source representations. We begin by collecting expressions for the delta function in cylindrical and spherical coordinates. We follow with a discussion of time-harmonic representations of functions and vectors. We next introduce the electromagnetic model in the time domain, and then specialize to the time-harmonic case. We begin our study of electromagnetic sources with a consideration of the sheet current source. We then, in sequence, study the line source, the ring source, and the point source. Throughout, we shall utilize the Green's function method and the spectral representation method, developed in Chapters 2 and 3, respectively, in order to obtain alternative representations of the fields from various sources.

4.2 DELTA FUNCTION TRANSFORMATIONS

In what follows, we shall require delta function representations in two and three dimensions. Depending on the particulars of the analysis, it will be convenient to express the results in different coordinate systems. In particular, we consider rectangular, polar, cylindrical, and spherical coordinates.

We begin with a point source in two dimensions (Fig. 4-1), located at the point $Q(x', y')$. We represent this point source by $\delta(x - x')\delta(y - y')$

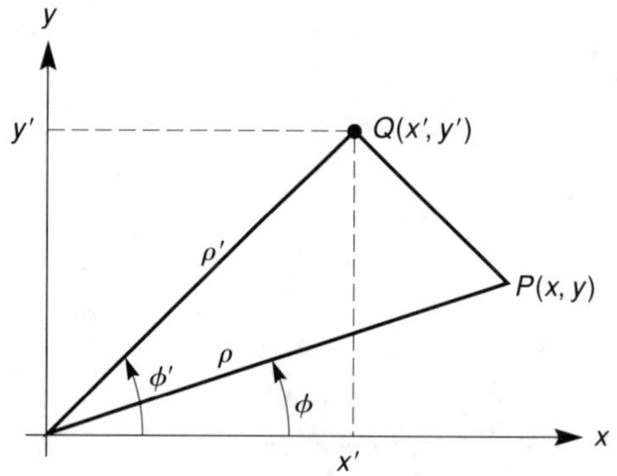

Fig. 4-1 Point source in two dimensions, located at (x', y').

and seek a corresponding representation in polar coordinates. Since the polar coordinate point corresponding to (x', y') is (ρ', ϕ'), we write

$$\delta(x - x')\delta(y - y') = f_1(\rho, \phi)\delta(\rho - \rho')\delta(\phi - \phi') \tag{4.1}$$

The function $f_1(\rho, \phi)$ allows for the possibility of an additional factor other than the delta functions that might be introduced by the Jacobian in the coordinate transformation. Integrating both sides of (4.1) over the xy-plane gives

$$1 = \int_0^{2\pi} \int_0^{\infty} f_1(\rho, \phi)\delta(\rho - \rho')\delta(\phi - \phi')\rho d\rho d\phi \tag{4.2}$$

Equation (4.2) is reduced to an identity by the choice

$$f_1(\rho, \phi) = \frac{1}{\rho} \tag{4.3}$$

Substitution into (4.1) gives

$$\delta(x - x')\delta(y - y') = \frac{\delta(\rho - \rho')\delta(\phi - \phi')}{\rho} \tag{4.4}$$

Equation (4.4) gives the polar representation of a point source in a plane as long as the source is at a location other than the origin. For a point source at the origin, we have $\delta(x)\delta(y)$. In transforming this representation to polar coordinates, we note that the origin in polar coordinates is given by $\rho = 0$, *independent of* ϕ. We say that the coordinate ϕ is *ignorable* at the origin [1] and write

$$\delta(x)\delta(y) = f_2(\rho)\delta(\rho) \tag{4.5}$$

Integrating (4.5) over the xy-plane gives

$$1 = \int_0^{2\pi} \int_0^{\infty} f_2(\rho)\delta(\rho)\rho d\rho d\phi$$

$$= \int_0^{\infty} [2\pi\rho f_2(\rho)]\delta(\rho)d\rho \tag{4.6}$$

Equation (4.6) is reduced to an identity by the choice

$$f_2(\rho) = \frac{1}{2\pi\rho} \tag{4.7}$$

Substitution into (4.5) gives

$$\delta(x)\delta(y) = \frac{\delta(\rho)}{2\pi\rho} \tag{4.8}$$

The proper transformations in three dimensions between rectangular and cylindrical coordinates require no further analysis since the z-coordinate remains the same in both systems. We have

$$\delta(x - x')\delta(y - y')\delta(z - z') = \frac{\delta(\rho - \rho')\delta(\phi - \phi')\delta(z - z')}{\rho} \tag{4.9}$$

$$\delta(x)\delta(y)\delta(z) = \frac{\delta(\rho)\delta(z)}{2\pi\rho} \tag{4.10}$$

In transformations to spherical coordinates, there are three cases of interest. We begin with a point source (Fig. 4-2) at the location $Q(x', y', z')$ and write

$$\delta(x-x')\delta(y-y')\delta(z-z') = f_3(r, \theta, \phi)\delta(r-r')\delta(\theta-\theta')\delta(\phi-\phi') \tag{4.11}$$

Integrating both sides over all space gives

$$1 = \int_0^{2\pi} \int_0^{\pi} \int_0^{\infty} f_3(r, \theta, \phi)\delta(r - r')\delta(\theta - \theta')\delta(\phi - \phi')r^2 \sin\theta dr d\theta d\phi \tag{4.12}$$

Equation (4.12) is reduced to an identity by the choice

$$f_3(r, \theta, \phi) = \frac{1}{r^2 \sin\theta} \tag{4.13}$$

Therefore,

$$\delta(x - x')\delta(y - y')\delta(z - z') = \frac{\delta(r - r')\delta(\theta - \theta')\delta(\phi - \phi')}{r^2 \sin\theta} \tag{4.14}$$

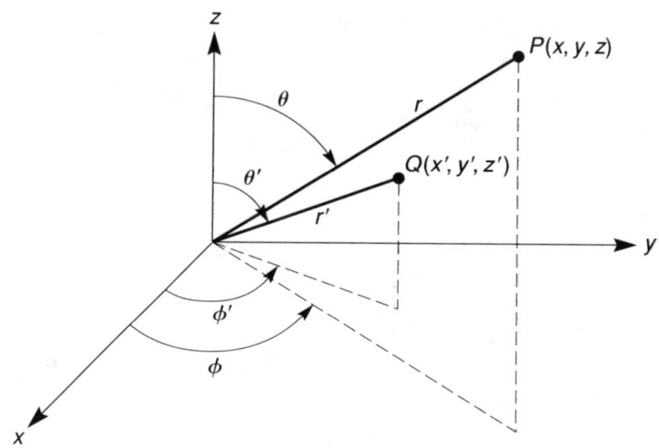

Fig. 4-2 Point source in three dimensions, located at (x', y', z').

In the case where the point source is located along the z-axis, the coordinate ϕ is ignorable. We have

$$\delta(x)\delta(y)\delta(z-z') = f_4(r,\theta)\delta(r-r')\delta(\theta) \tag{4.15}$$

Integrating over all space gives

$$\begin{aligned} 1 &= \int_0^{2\pi}\int_0^{\pi}\int_0^{\infty} f_4(r,\theta)\delta(r-r')\delta(\theta)r^2\sin\theta\, dr\, d\theta\, d\phi \\ &= \int_0^{\pi}\int_0^{\infty} [2\pi r^2 \sin\theta f_4(r,\theta)]\delta(r-r')\delta(\theta)\, dr\, d\theta \end{aligned} \tag{4.16}$$

Equation (4.16) is reduced to an identity by the choice

$$f_4(r,\theta) = \frac{1}{2\pi r^2 \sin\theta} \tag{4.17}$$

Substitution into (4.15) gives

$$\delta(x)\delta(y)\delta(z-z') = \frac{\delta(r-r')\delta(\theta)}{2\pi r^2\sin\theta} \tag{4.18}$$

In the case where the point source is at the origin (0, 0, 0), the coordinates θ and ϕ are both ignorable. We have

$$\delta(x)\delta(y)\delta(z) = f_5(r)\delta(r) \tag{4.19}$$

Integrating over all space gives

$$\begin{aligned} 1 &= \int_0^{2\pi}\int_0^{\pi}\int_0^{\infty} f_5(r)\delta(r)r^2\sin\theta\, dr\, d\theta\, d\phi \\ &= \int_0^{\infty} [4\pi r^2 f_5(r)]\delta(r)\, dr \end{aligned} \tag{4.20}$$

Equation (4.20) is reduced to an identity by the choice

$$f_5(r) = \frac{1}{4\pi r^2} \tag{4.21}$$

Substitution into (4.19) gives

$$\delta(x)\delta(y)\delta(z) = \frac{\delta(r)}{4\pi r^2} \tag{4.22}$$

4.3 TIME-HARMONIC REPRESENTATIONS

In subsequent considerations of the electromagnetic model, we shall be dealing with quantities that vary harmonically with time t. The time-harmonic representation is useful in the determination of the cosinusoidal steady-state behavior of the electromagnetic fields. The representation can also be directly extended to give the response for more general forms of source input [2]. An excellent treatment of time-harmonic representations is given in [3]. We shall include herein only the major results. A time-harmonic function $f(t)$ has the form

$$f(t) = F_0 \cos(\omega t + \phi) \tag{4.23}$$

where ω is radian frequency and where F_0 and ϕ are real and time-independent. We write this function in terms of the real-part operator as follows:

$$f(t) = \mathrm{Re}\left(Fe^{i\omega t}\right) \tag{4.24}$$

where

$$F = F_0 e^{i\phi} \tag{4.25}$$

That (4.24) is equivalent to (4.23) can be demonstrated by substituting (4.25) into (4.24) and performing the real-part operation. The details are left to the reader. We may show the following relation for derivatives:

$$\frac{df}{dt} = \mathrm{Re}(i\omega F e^{i\omega t}) \tag{4.26}$$

The proof is straightforward and is left for the problems.

The real-part operator has the following useful properties [4]: For $z, z_1, z_2 \in \mathbf{C}$ and $a, t \in \mathbf{R}$,

$$\mathrm{Re}(z_1) + \mathrm{Re}(z_2) = \mathrm{Re}(z_1 + z_2) \tag{4.27}$$

$$\mathrm{Re}(az) = a\mathrm{Re}(z) \tag{4.28}$$

$$\frac{\partial}{\partial t}[\text{Re}(z)] = \text{Re}\left(\frac{\partial z}{\partial t}\right) \tag{4.29}$$

$$\int \text{Re}(z)dt = \text{Re}\left(\int z dt\right) \tag{4.30}$$

These real-part relationships are easily verified. The details are left for the problems. An additional relationship involving the real-part operator is the following: If $A, B \in \mathbf{C}$ and are time-independent, and if

$$\text{Re}(Ae^{i\omega t}) = \text{Re}(Be^{i\omega t}), \qquad \text{all } t \tag{4.31}$$

then

$$A = B \tag{4.32}$$

Indeed, let $t = 0$. Then $\text{Re}(A) = \text{Re}(B)$. Let $\omega t = \pi/2$. Then $\text{Re}(iA) = \text{Re}(iB)$, which implies that $\text{Im}(A) = \text{Im}(B)$. Since A and B are time-independent, the above equality of their real parts and their imaginary parts must be true for all t, and thus $A = B$.

The complex representation of time-harmonic functions can be extended to vectors. Indeed, let $\mathcal{E}(t)$ be a real, time-varying vector with time-harmonic components, viz.

$$\mathcal{E}(t) = \hat{x}E_x\cos(\omega t+\phi_x)+\hat{y}E_y\cos(\omega t+\phi_y)+\hat{z}E_z\cos(\omega t+\phi_z) \tag{4.33}$$

where $\hat{x}, \hat{y}, \hat{z}$ are unit vectors in the three Cartesian coordinate directions and $E_x, E_y, E_z, \phi_x, \phi_y, \phi_z$ are real and time-independent. In the same manner as in the above treatment of complex functions, complex vectors can be written in terms of the real-part operator, as follows:

$$\mathcal{E}(t) = \text{Re}(\mathbf{E}e^{i\omega t}) \tag{4.34}$$

where

$$\mathbf{E} = \hat{x}E_x e^{i\phi_x} + \hat{y}E_y e^{i\phi_y} + \hat{z}E_z e^{i\phi_z} \tag{4.35}$$

The equivalence of (4.34) and (4.33) can be verified by substituting (4.35) into (4.34) and performing the real-part operation. The details are left for the reader.

4.4 THE ELECTROMAGNETIC MODEL

The macroscopic model for the behavior of electromagnetic fields is given by the following set of equations:

$$\nabla \times \mathcal{E} = -\frac{\partial \mathcal{B}}{\partial t} - \mathcal{M} \tag{4.36}$$

$$\nabla \times \mathcal{H} = \frac{\partial \mathcal{D}}{\partial t} + \mathcal{J} \tag{4.37}$$

$$\nabla \cdot \mathcal{D} = \rho \tag{4.38}$$

$$\nabla \cdot \mathcal{B} = \rho_m \tag{4.39}$$

$$\nabla \cdot \mathcal{J} = -\frac{\partial \rho}{\partial t} \tag{4.40}$$

$$\nabla \cdot \mathcal{M} = -\frac{\partial \rho_m}{\partial t} \tag{4.41}$$

where the symbols are defined as follows:

$\mathcal{E}$ electric field intensity (volts/meter)
$\mathcal{H}$ magnetic field intensity (amps/meter)
$\mathcal{D}$ electric flux density (coulombs/meter^2)
$\mathcal{B}$ magnetic flux density (webers/meter^2)
$\mathcal{J}$ electric current density (amps/meter^2)
$\mathcal{M}$ magnetic current density (volts/meter^2)
ρ electric charge density (coulombs/meter^3)
ρ_m magnetic charge density (webers/meter^3)

In performing dimensional analyses, it is helpful to have the following equalities:

coulomb=amp · second
weber=volt · second
ohm=volt/amp

All field and source quantities vary with both space and time. Typically, for the electric field, we have $\mathcal{E}(x, y, z, t)$, which we write in shorthand notation as $\mathcal{E}(\mathbf{r}, t)$. The magnetic charge $\rho_m(\mathbf{r}, t)$ and the magnetic current $\mathcal{M}(\mathbf{r}, t)$ have not been shown to exist in nature, but can be defined as *equivalent* sources in equivalence theorems involving the electric fields [5]. We shall assume that all quantities vary time-harmonically. Typically,

$$\mathcal{E}(\mathbf{r}, t) = \text{Re}\left[\mathbf{E}(\mathbf{r}, \omega)e^{i\omega t}\right] \tag{4.42}$$

Then, in (4.36), we have

$$\nabla \times \left[\text{Re}(\mathbf{E}e^{i\omega t})\right] = -\frac{\partial}{\partial t}\left[\text{Re}(\mathbf{B}e^{i\omega t})\right] - \text{Re}(\mathbf{M}e^{i\omega t}) \tag{4.43}$$

Using the real-part operator relationships in (4.26)–(4.29), we obtain

$$\mathrm{Re}\left[\nabla \times (\mathbf{E}e^{i\omega t})\right] = -\mathrm{Re}\left[(i\omega\mathbf{B} + \mathbf{M})e^{i\omega t}\right] \tag{4.44}$$

But, by a well-known vector identity,

$$\nabla \times (\mathbf{E}e^{i\omega t}) = (\nabla \times \mathbf{E})e^{i\omega t} \tag{4.45}$$

Substituting (4.45) into (4.44) and applying (4.31) and (4.32), we obtain

$$\nabla \times \mathbf{E} = -i\omega\mathbf{B} - \mathbf{M} \tag{4.46}$$

A similar procedure in (4.37)–(4.41) yields

$$\nabla \times \mathbf{H} = i\omega\mathbf{D} + \mathbf{J} \tag{4.47}$$

$$\nabla \cdot \mathbf{D} = \rho \tag{4.48}$$

$$\nabla \cdot \mathbf{B} = \rho_m \tag{4.49}$$

$$\nabla \cdot \mathbf{J} = -i\omega\rho \tag{4.50}$$

$$\nabla \cdot \mathbf{M} = -i\omega\rho_m \tag{4.51}$$

We remark that, in keeping with usual practice, we have used the same symbols for electric charge density in both the time and frequency domains, and similarly for magnetic charge density. Which domain we are considering will be clear in context.

In a *simple medium*, the electric flux density **D** is simply related to the electric field intensity **E** by

$$\mathbf{D} = \epsilon\mathbf{E} \tag{4.52}$$

where ϵ is the permittivity of the medium in farads/meter. Similarly, for the magnetic field,

$$\mathbf{B} = \mu\mathbf{H} \tag{4.53}$$

where μ is the permeability of the medium in henrys/meter. The basic units of farads can be deduced from the units of **D** and **E** in (4.52), and similarly for the units of henrys in (4.53).

Equations (4.46)–(4.53) constitute the *electromagnetic model* in simple media. We shall use the model in what follows to develop representations of electromagnetic sources. In Chapter 5, we shall use the model to develop solutions to some example boundary value problems.

4.5 THE SHEET CURRENT SOURCE

Consider a planar electric current sheet in the xy-plane (Fig. 4-3). We assume linear polarization in the x-direction and no variations in either x or y. The mathematical representation of this current sheet involves the electric current density $\mathbf{J}(z)$ in amps/m^2, given by

$$\mathbf{J}(z) = \hat{x} J_{s0}\delta(z) \tag{4.54}$$

where $\hat{x}$ is a unit vector in the x-direction and J_{s0} is a constant surface current density in amps/m. We write Maxwell's curl equations by substituting (4.52) into (4.47) and (4.53) into (4.46) to give

$$\nabla \times \mathbf{E} = -i\omega\mu\mathbf{H} - \mathbf{M} \tag{4.55}$$

$$\nabla \times \mathbf{H} = i\omega\epsilon\mathbf{E} + \mathbf{J} \tag{4.56}$$

In general, the permittivity is complex and is given by [6]

$$\epsilon = \epsilon_d + \frac{\sigma}{i\omega} \tag{4.57}$$

where σ is the conductivity in mhos/m and ϵ_d is the permittivity of a perfect dielectric ($\sigma = 0$). The complex permittivity accounts for losses in the medium. For the case where the losses are negligible, the complex permittivity reduces to the perfect dielectric permittivity ϵ_d.

Since the source, given by (4.54), varies only with z, and since there are no scattering objects present, we conclude that $\partial/\partial x = \partial/\partial y = 0$. Substituting this result and (4.54) into Maxwell's equations, setting $\mathbf{M} = 0$, and expanding in Cartesian coordinates, we obtain

$$-\frac{dH_y}{dz} = J_{s0}\delta(z) + i\omega\epsilon E_x \tag{4.58}$$

$$\frac{dH_x}{dz} = i\omega\epsilon E_y \tag{4.59}$$

$$0 = i\omega\epsilon E_z \tag{4.60}$$

$$\frac{dE_y}{dz} = i\omega\mu H_x \tag{4.61}$$

$$\frac{dE_x}{dz} = -i\omega\mu H_y \tag{4.62}$$

$$0 = i\omega\mu H_z \tag{4.63}$$

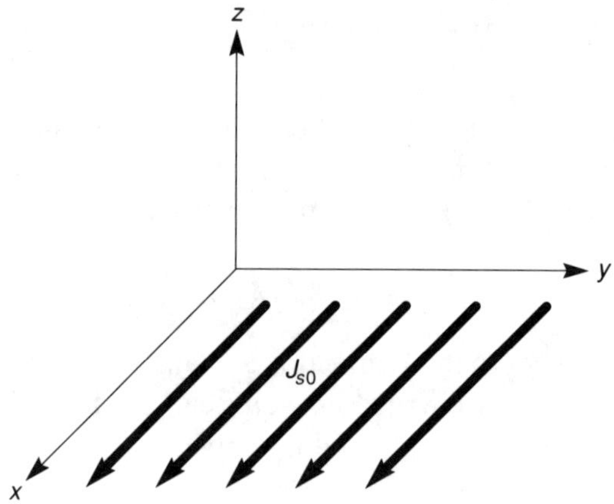

Fig. 4-3 Electric current sheet located in the xy-plane and linearly polarized in the x-direction. Current sheet extends over entire xy-plane.

From (4.60) and (4.63), we find that

$$E_z = H_z = 0 \tag{4.64}$$

The remaining equations decouple into two independent sets. The first set is given by

$$\frac{dE_x}{dz} = -i\omega\mu H_y \tag{4.65}$$

$$-\frac{dH_y}{dz} = J_{s0}\delta(z) + i\omega\epsilon E_x \tag{4.66}$$

The second set is given by

$$\frac{dE_y}{dz} = i\omega\mu H_x \tag{4.67}$$

$$\frac{dH_x}{dz} = i\omega\epsilon E_y \tag{4.68}$$

Note that the first set contains E_x, H_y, and J_{s0}, while the second set contains E_y, H_x, and no sources. Since the second set is source-free throughout all space and is not coupled in any manner to the first set, we must conclude that the only solution to (4.67) and (4.68) is the trivial solution, viz.

$$E_y = H_x = 0 \tag{4.69}$$

The problem is therefore completely described by (4.65) and (4.66), together with appropriate conditions as $z \to \pm\infty$. We take the derivative of (4.65) with respect to z and substitute (4.66) to obtain the following set:

$$\frac{d^2 E_x}{dz^2} + k^2 E_x = i\omega\mu J_{s0}\delta(z) \tag{4.70}$$

$$H_y = -\frac{1}{i\omega\mu}\frac{dE_x}{dz} \tag{4.71}$$

where the wavenumber k is given by

$$k = k_d\sqrt{1 - iS} \tag{4.72}$$

with

$$S = \frac{\sigma}{\omega\epsilon_d} \tag{4.73}$$

and

$$k_d = \omega\sqrt{\mu\epsilon_d} \tag{4.74}$$

The wavenumber k_d is the wavenumber that would be present in a perfect dielectric ($\sigma = 0$). In the engineering literature, S is called the *loss tangent* [7]. We note that, if the second-order, linear, ordinary differential equation in (4.70) can be solved for the electric field E_x, the magnetic field H_y can be found by the simple differentiation indicated in (4.71). For limiting conditions, we demand that, for $k \in \mathbf{C}$,

$$\lim_{z\to\pm\infty} E_x = 0 \tag{4.75}$$

To solve the differential equation in (4.70), we let

$$E_x = -i\omega\mu J_{s0} g \tag{4.76}$$

Substitution into (4.70) yields

$$-\frac{d^2 g}{dz^2} - k^2 g = \delta(z) \tag{4.77}$$

with limiting conditions

$$\lim_{z\to\pm\infty} g = 0 \tag{4.78}$$

We recognize this as a Green's function problem in SLP3. In general, the Green's function is associated with the delta-function source $\delta(z - \zeta)$.

This source would produce a Green's function $g(z, \zeta)$. In the case in (4.77), $\zeta = 0$ and we obtain $g(z, 0)$. The solution to this Green's function problem has been given in Example 2.20, viz.

$$g(z, 0) = \frac{e^{-ik|z|}}{2ik}, \qquad \text{Im}(k) < 0 \tag{4.79}$$

Substitution into (4.76) gives

$$E_x(z) = -\frac{\omega\mu}{2k} J_{s0} e^{-ik|z|} \tag{4.80}$$

We may normalize this result by letting

$$J_{s0} = -\frac{2k}{\omega\mu} \tag{4.81}$$

We then have

$$E_x(z) = e^{-ik|z|} \tag{4.82}$$

Substitution of this result into (4.71) yields the accompanying magnetic field, viz.

$$H_y(z) = \frac{1}{\eta} \begin{cases} e^{-ikz}, & z > 0 \\ -e^{ikz}, & z < 0 \end{cases} \tag{4.83}$$

where the intrinsic impedance η is given by

$$\eta = \frac{\omega\mu}{k} \tag{4.84}$$

The solution to (4.77) with limiting conditions in (4.78) has been obtained by the Green's function method. Alternately, the solution can also be obtained by spectral methods. From the result in Example 3.4, the spectral representation of the delta function for the operator $-d^2/dz^2$ with limiting conditions given by (4.78) is given by

$$\delta(z - \zeta) = \frac{1}{2\pi} \int_{-\infty}^{\infty} e^{i\beta(z-\zeta)} d\beta \tag{4.85}$$

This spectral representation defines the Fourier transform. Applying this transform to the Green's function, we have

$$g(z, 0) = \frac{1}{2\pi} \int_{-\infty}^{\infty} G(\beta, 0) e^{i\beta z} d\beta \tag{4.86}$$

$$G(\beta, 0) = \int_{-\infty}^{\infty} g(z, 0)e^{-i\beta z} dz \tag{4.87}$$

Taking the Fourier transform of (4.77) and rearranging gives

$$G(\beta, 0) = \frac{1}{\beta^2 - k^2} \tag{4.88}$$

Taking the inverse Fourier transform yields the alternative solution form

$$g(z, 0) = \frac{1}{2\pi} \int_{-\infty}^{\infty} \frac{e^{i\beta z}}{\beta^2 - k^2} d\beta \tag{4.89}$$

Substituting into (4.76) and applying (4.81), we find that

$$E_x(z) = -\frac{ik}{\pi} \int_{-\infty}^{\infty} \frac{e^{i\beta z}}{k^2 - \beta^2} d\beta \tag{4.90}$$

Since the solution to (4.70) is unique, we may equate the two results in (4.82) and (4.90) to give the following useful relationship:

$$e^{-ik|z|} = \frac{ik}{\pi} \int_{-\infty}^{\infty} \frac{e^{i\beta z}}{\beta^2 - k^2} d\beta, \qquad \text{Im}(k) < 0 \tag{4.91}$$

We remark that the result in (4.91) can also be obtained by contour integration techniques, as we show in the following example.

EXAMPLE 4.1 We shall show that

$$\int_{-\infty}^{\infty} \frac{e^{i\beta z}}{\beta^2 - k^2} d\beta = \frac{\pi}{ik} e^{-ik|z|}, \qquad \text{Im}(k) < 0$$

Consider

$$\oint_{C_R + C_1} \frac{e^{i\beta z}}{\beta^2 - k^2} d\beta \tag{4.92}$$

around a closed contour (Fig. 4-4) along the real axis from $-R$ to R and a semicircle of radius R through the upper half-plane. We constrain R such that $R > |k|$. The denominator of (4.92) has simple poles at $\beta = \pm k$. We have $\text{Im}(k) < 0$. It can be shown that this selection of the sign of the imaginary part, together with the definition of k in (4.72), implies that $\text{Re}(k) > 0$. The details are left for the problems. Therefore, the pole at $\beta = -k$ is enclosed by the contour, and the residue theorem gives

$$\oint_{C_R + C_1} \frac{e^{i\beta z}}{\beta^2 - k^2} d\beta = 2\pi i \text{Res}\left[\frac{e^{i\beta z}}{\beta^2 - k^2}; -k\right] \tag{4.93}$$

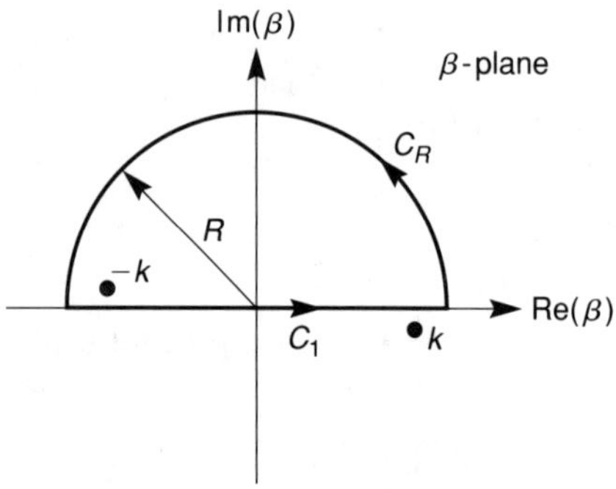

Fig. 4-4 Contour for the evaluation of the contour integral in (4.92).

Evaluating the residue and splitting the contour integral into two pieces gives

$$\int_{-R}^{R} \frac{e^{i\beta z}}{\beta^2 - k^2} d\beta + \int_{C_R} \frac{e^{i\beta z}}{\beta^2 - k^2} d\beta = \frac{\pi}{ik} e^{-ikz} \tag{4.94}$$

We now show that

$$\lim_{R \to \infty} \left| \int_{C_R} \frac{e^{i\beta z}}{\beta^2 - k^2} d\beta \right| = 0, \qquad z > 0 \tag{4.95}$$

Indeed, on C_R, let

$$\beta = Re^{i\theta} \tag{4.96}$$

Then,

$$\begin{aligned} \left| \int_{C_R} \frac{e^{i\beta z}}{\beta^2 - k^2} d\beta \right| &= \left| \int_0^{\pi} \frac{e^{izR\cos\theta} e^{-zR\sin\theta} i Re^{i\theta} d\theta}{R^2 e^{i2\theta} - k^2} \right| \\ &\leq \frac{R}{R^2 - |k|^2} \int_0^{\pi} e^{-zR\sin\theta} d\theta \\ &\leq \frac{R\pi}{R^2 - |k|^2} \end{aligned} \tag{4.97}$$

In the last inequality, we have used the fact that $zR\sin\theta > 0, 0 < \theta < \pi$ to bound the integrand with unity. Therefore, in the limit as $R \to \infty$, the integral around C_R approaches zero. Taking the limit in (4.94) gives

$$\int_{-\infty}^{\infty} \frac{e^{i\beta z}}{\beta^2 - k^2} d\beta = \frac{\pi}{ik} e^{-ikz}, \qquad z > 0 \tag{4.98}$$

Equation (4.98) gives the result for $z > 0$. For $z < 0$, we close the contour C_1 along a semi-circle through the lower half-plane. The details are left for the problems.

■

4.6 THE LINE SOURCE

Consider a line current source located along the z-axis (Fig. 4-5) and extending from $z = -\infty$ to $z = \infty$. We represent the current density associated with this source by

$$\mathbf{J}(\rho) = \hat{z} I_0 \delta(x)\delta(y) \tag{4.99}$$

where I_0 is a constant current in amps. We begin our study of the fields produced by this current source by considering the problem in cylindrical coordinates. From (4.8), the cylindrical coordinate representation of the current $\mathbf{J}(\rho)$ is given by

$$\mathbf{J}(\rho) = \hat{z} I_0 \frac{\delta(\rho)}{2\pi\rho} \tag{4.100}$$

Since the current source is independent of ϕ and z, and since there are no scattering objects, we must have $\partial/\partial\phi = \partial/\partial z = 0$. Maxwell's curl equations in (4.55) and (4.56) can therefore be written in cylindrical coordinates by

$$E_\rho = H_\rho = 0 \tag{4.101}$$

$$\frac{dH_z}{d\rho} = -i\omega\epsilon E_\phi \tag{4.102}$$

$$\frac{1}{\rho}\left[\frac{d}{d\rho}(\rho E_\phi)\right] = -i\omega\mu H_z \tag{4.103}$$

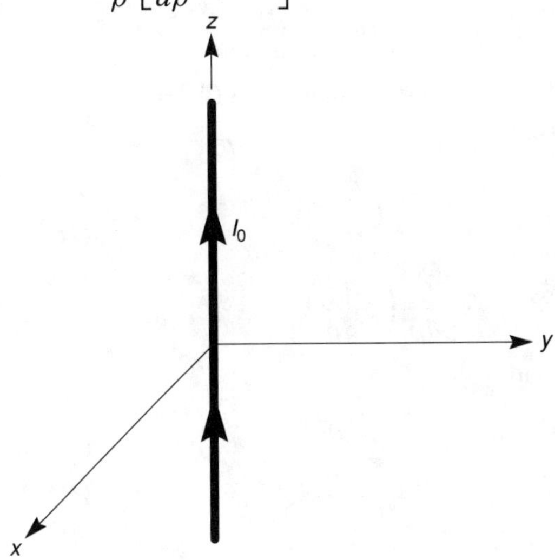

Fig. 4-5 Electric line current located along the z-axis and extending from $z = -\infty$ to $z = \infty$.

$$\frac{dE_z}{d\rho} = i\omega\mu H_\phi \tag{4.104}$$

$$\frac{1}{\rho}\left[\frac{d}{d\rho}(\rho H_\phi)\right] = J_z + i\omega\epsilon E_z \tag{4.105}$$

where

$$J_z = I_0 \frac{\delta(\rho)}{2\pi\rho} \tag{4.106}$$

We note that (4.102) and (4.103) are source-free and independent of (4.104) and (4.105). Therefore,

$$H_z = E_\phi = 0 \tag{4.107}$$

We conclude that (4.104) and (4.105), together with appropriate boundary and/or limiting conditions, completely characterize the problem. We multiply (4.104) by ρ, take the derivative with respect to ρ, multiply by $1/\rho$, and then divide by $i\omega\mu$ to obtain

$$\frac{1}{i\omega\mu\rho}\left[\frac{d}{d\rho}\left(\rho\frac{dE_z}{d\rho}\right)\right] = \frac{1}{\rho}\left[\frac{d}{d\rho}(\rho H_\phi)\right] \tag{4.108}$$

Substitution of (4.108) into (4.105) produces the following set:

$$\frac{1}{\rho}\left[\frac{d}{d\rho}\left(\rho\frac{dE_z}{d\rho}\right)\right] + k^2 E_z = i\omega\mu I_0 \frac{\delta(\rho)}{2\pi\rho} \tag{4.109}$$

$$H_\phi = \frac{1}{i\omega\mu}\frac{dE_z}{d\rho} \tag{4.110}$$

To solve the differential equation in (4.109), we let

$$g = -\frac{2\pi E_z}{i\omega\mu I_0} \tag{4.111}$$

and obtain

$$\frac{1}{\rho}\left[\frac{d}{d\rho}\left(\rho\frac{dg}{d\rho}\right)\right] + k^2 g = -\frac{\delta(\rho)}{\rho} \tag{4.112}$$

with limiting condition

$$\lim_{\rho\to\infty} g = 0 \tag{4.113}$$

To solve (4.112), we first consider a result from Example 2.21. In that example, we considered

$$\frac{1}{\rho}\left[\frac{d}{d\rho}\left(\rho\frac{dg}{d\rho}\right)\right] + k^2 g = -\frac{\delta(\rho-\rho')}{\rho} \tag{4.114}$$

with the limiting condition given in (4.113) and a finiteness condition at the origin. The solution was given in (2.184) and is repeated here with some trivial changes in notation, viz.

$$g(\rho, \rho') = \frac{\pi}{2i} \begin{cases} H_0^{(2)}(k\rho') J_0(k\rho), & \rho < \rho' \\ H_0^{(2)}(k\rho) J_0(k\rho'), & \rho > \rho' \end{cases} \tag{4.115}$$

Taking the limit as $\rho' \to 0$ yields the solution to (4.112), viz.

$$g(\rho, 0) = \frac{\pi}{2i} H_0^{(2)}(k\rho) \tag{4.116}$$

Substituting (4.116) into (4.111) and solving for E_z, we obtain

$$E_z = -\frac{\omega\mu I_0}{4} H_0^{(2)}(k\rho) \tag{4.117}$$

Substitution of this result into (4.110) gives

$$H_\phi = -\frac{ikI_0}{4} H_1^{(2)}(k\rho) \tag{4.118}$$

The solution to (4.114) with limiting condition (4.113) at infinity and a finiteness condition at the origin has been obtained by the Green's function method. Alternately, the solution can also be obtained by spectral methods. From Example 3.5, the spectral representation of the differential operator in (4.114) with the given limiting and finiteness conditions is given by

$$\frac{\delta(\rho - \rho')}{\rho} = \int_0^\infty J_0(\lambda\rho) J_0(\lambda\rho') \lambda d\lambda \tag{4.119}$$

As found in Example 3.5, this representation gives the Fourier–Bessel transform pair

$$F(\lambda) = \int_0^\infty f(\rho) J_0(\lambda\rho) \rho d\rho \tag{4.120}$$

$$f(\rho) = \int_0^\infty F(\lambda) J_0(\lambda\rho) \lambda d\lambda \tag{4.121}$$

Taking the Fourier–Bessel transform of both sides of (4.114) gives

$$(-\lambda^2 + k^2) G(\lambda, \rho') = -J_0(\lambda\rho') \tag{4.122}$$

where $G(\lambda, \rho')$ is the Fourier–Bessel transform of $g(\rho, \rho')$. Solving for G and taking the inverse Fourier–Bessel transform gives

$$g(\rho, \rho') = \int_0^\infty \frac{J_0(\lambda\rho) J_0(\lambda\rho')}{\lambda^2 - k^2} \lambda d\lambda \tag{4.123}$$

To produce the solution to (4.112), we take the limit as $\rho' \to 0$ and obtain

$$g(\rho, 0) = \int_0^\infty \frac{J_0(\lambda\rho)}{\lambda^2 - k^2} \lambda d\lambda \tag{4.124}$$

Substitution into (4.111) yields

$$E_z = -\frac{i\omega\mu I_0}{2\pi} \int_0^\infty \frac{J_0(\lambda\rho)}{\lambda^2 - k^2} \lambda d\lambda \tag{4.125}$$

Comparing (4.125) to (4.117), we obtain the following integral representation of the Hankel function [8]:

$$H_0^{(2)}(k\rho) = \frac{2i}{\pi} \int_0^\infty \frac{J_0(\lambda\rho)}{\lambda^2 - k^2} \lambda d\lambda \tag{4.126}$$

Taking the Fourier–Bessel transform gives the relation

$$\frac{1}{\lambda^2 - k^2} = \frac{\pi}{2i} \int_0^\infty H_0^{(2)}(k\rho) J_0(\lambda\rho) \rho d\rho \tag{4.127}$$

We have obtained two representations of the electric field E_z produced by a line current located along the z-axis. These representations are given by (4.117) and (4.125). Further representations can be obtained by considering the same problem in Cartesian coordinates. Since $\partial/\partial z = 0$, Maxwell's curl equations reduce to

$$\frac{\partial E_z}{\partial y} = -i\omega\mu H_x \tag{4.128}$$

$$\frac{\partial E_z}{\partial x} = i\omega\mu H_y \tag{4.129}$$

$$\frac{\partial E_y}{\partial x} - \frac{\partial E_x}{\partial y} = -i\omega\mu H_z \tag{4.130}$$

$$\frac{\partial H_z}{\partial y} = i\omega\epsilon E_x \tag{4.131}$$

$$\frac{\partial H_z}{\partial x} = -i\omega\epsilon E_y \tag{4.132}$$

$$\frac{\partial H_y}{\partial x} - \frac{\partial H_x}{\partial y} = I_0\delta(x)\delta(y) + i\omega\epsilon E_z \tag{4.133}$$

These six equations can be grouped into two independent sets, as follows:

Set 1: TM_z

$$\frac{\partial E_z}{\partial y} = -i\omega\mu H_x \tag{4.134}$$

$$\frac{\partial E_z}{\partial x} = i\omega\mu H_y \tag{4.135}$$

$$\frac{\partial H_y}{\partial x} - \frac{\partial H_x}{\partial y} = I_0\delta(x)\delta(y) + i\omega\epsilon E_z \tag{4.136}$$

Set 2: TE_z

$$\frac{\partial H_z}{\partial y} = i\omega\epsilon E_x \tag{4.137}$$

$$\frac{\partial H_z}{\partial x} = -i\omega\epsilon E_y \tag{4.138}$$

$$\frac{\partial E_y}{\partial x} - \frac{\partial E_x}{\partial y} = -i\omega\mu H_z \tag{4.139}$$

Set 1 is labeled TM_z since it contains no magnetic field component in the z-direction. In a similar manner, Set 2 is TE_z. We note that the two sets are not coupled and that Set 2 is source-free. Therefore, the only solution to Set 2 consists of the null fields, viz.

$$H_z = E_x = E_y = 0 \tag{4.140}$$

To solve for the TM_z fields, we differentiate (4.134) with respect to y, (4.135) with respect to x, add the result, and substitute (4.136) to obtain

$$\frac{\partial^2 E_z}{\partial x^2} + \frac{\partial^2 E_z}{\partial y^2} + k^2 E_z = i\omega\mu I_0\delta(x)\delta(y) \tag{4.141}$$

$$H_x = -\frac{1}{i\omega\mu}\frac{\partial E_z}{\partial y} \tag{4.142}$$

$$H_y = \frac{1}{i\omega\mu}\frac{\partial E_z}{\partial x} \tag{4.143}$$

We consider (4.141), together with the limiting conditions

$$\lim_{x\to\pm\infty} E_z(x, y) = 0 \tag{4.144}$$

$$\lim_{y\to\pm\infty} E_z(x, y) = 0 \tag{4.145}$$

To reduce (4.141), we combine the spectral representation and Green's function methods. First, from Example 3.4, the spectral representation of $\partial^2/\partial x^2$ with limiting condition in (4.144) is given by

$$\delta(x - x') = \frac{1}{2\pi} \int_{-\infty}^{\infty} e^{ik_x(x-x')} dk_x \tag{4.146}$$

Multiplying both sides of (4.146) by $E_z(x', y)$ and integrating over $(-\infty, \infty)$ gives the *spatial Fourier transform* pair

$$E_z(x, y) = \frac{1}{2\pi} \int_{-\infty}^{\infty} \hat{E}_z(k_x, y) e^{ik_x x} dk_x \tag{4.147}$$

$$\hat{E}_z(k_x, y) = \int_{-\infty}^{\infty} E_z(x, y) e^{-ik_x x} dx \tag{4.148}$$

We therefore take the Fourier transform of both sides of (4.141) and produce

$$\frac{d^2 \hat{E}_z}{dy^2} + k_y^2 \hat{E}_z = i\omega\mu I_0 \delta(y) \tag{4.149}$$

where

$$k_y = \sqrt{k^2 - k_x^2} \tag{4.150}$$

and where $E_z(x, y)$ and $\hat{E}_z(k_x, y)$ are Fourier transform pairs. We let

$$G_1 = -\frac{\hat{E}_z}{i\omega\mu I_0} \tag{4.151}$$

so that

$$-\left(\frac{d^2}{dy^2} + k_y^2\right) G_1 = \delta(y) \tag{4.152}$$

where

$$\lim_{y \to \pm\infty} G_1(k_x, y) = 0 \tag{4.153}$$

We have previously produced the solution to this Green's function problem in (4.79). In this case, we have

$$G_1 = \frac{e^{-ik_y|y|}}{2ik_y}, \qquad \text{Im}(k_y) < 0 \tag{4.154}$$

Substituting into (4.151), solving for $\hat{E}_z$, and taking the inverse Fourier transform, we obtain

$$E_z(x, y) = -\frac{\omega\mu I_0}{4\pi} \int_{-\infty}^{\infty} \frac{e^{-ik_y|y|}}{k_y} e^{ik_x x} dk_x \tag{4.155}$$

Equation (4.155) gives another form of solution to the line source problem. If we compare (4.117) and (4.155), we produce the following integral representation of the Hankel function:

$$H_0^{(2)}\left[k(x^2+y^2)^{1/2}\right] = \frac{1}{\pi}\int_{-\infty}^{\infty} \frac{e^{-i(k^2-k_x^2)^{1/2}|y|}}{(k^2-k_x^2)^{1/2}} e^{ik_x x}\,dk_x \tag{4.156}$$

where

$$\mathrm{Im}(k^2-k_x^2)^{1/2} < 0 \tag{4.157}$$

Taking the transform of both sides of (4.156) yields

$$\frac{e^{-i(k^2-k_x^2)^{1/2}|y|}}{(k^2-k_x^2)^{1/2}} = \frac{1}{2}\int_{-\infty}^{\infty} H_0^{(2)}\left[k(x^2+y^2)^{1/2}\right] e^{-ik_x x}\,dx \tag{4.158}$$

We note in (4.141), (4.144), and (4.145) that the x and y differential operators and their manifolds are identical. We therefore could have taken the Fourier transform with respect to y. The result can be immediately obtained by interchanging x with y and k_x with k_y.

We now have three representations of the electric field from the line current source along the z-axis, given by (4.117), (4.125), and (4.155). A fourth representation can be obtained by taking the Fourier transforms of (4.141) with respect to x *and* y to obtain

$$\left(-k_x^2 - k_y^2 + k^2\right)\tilde{E}_z = i\omega\mu I_0 \tag{4.159}$$

where $E_z(x, y)$ and $\tilde{E}_z(k_x, k_y)$ are two-dimensional Fourier transform pairs. Solving for $\tilde{E}_z$ and taking the two-dimensional inverse transform gives

$$E_z(x,y) = i\omega\mu I_0 \left(\frac{1}{2\pi}\right)^2 \int_{-\infty}^{\infty}\int_{-\infty}^{\infty} \frac{e^{i(k_x x + k_y y)}}{k^2-k_x^2-k_y^2}\,dk_x dk_y \tag{4.160}$$

Comparing (4.117) and (4.160), we obtain the following double-integral representation of the Hankel function:

$$H_0^{(2)}\left[k(x^2+y^2)^{1/2}\right] = \frac{1}{i\pi^2}\int_{-\infty}^{\infty}\int_{-\infty}^{\infty} \frac{e^{i(k_x x + k_y y)}}{k^2-k_x^2-k_y^2}\,dk_x dk_y \tag{4.161}$$

Taking the two-dimensional transform, we obtain

$$\frac{1}{k^2-k_x^2-k_y^2} = \frac{i}{4}\int_{-\infty}^{\infty}\int_{-\infty}^{\infty} H_0^{(2)}\left[k(x^2+y^2)^{1/2}\right] e^{-i(k_x x + k_y y)}\,dx dy \tag{4.162}$$

We next move the line source away from the z-axis (Fig. 4-6) by defining a current

$$\mathbf{J} = \hat{z} I_0 \delta(x - x')\delta(y - y')$$

$$= \hat{z} I_0 \frac{\delta(\rho - \rho')}{\rho} \delta(\phi - \phi') \tag{4.163}$$

where we have used (4.4) to make the delta function coordinate transformation. We expand Maxwell's curl equations in cylindrical coordinates and obtain

$$\frac{1}{\rho}\frac{\partial H_z}{\partial \phi} = i\omega\epsilon E_\rho \tag{4.164}$$

$$\frac{\partial H_z}{\partial \rho} = -i\omega\epsilon E_\phi \tag{4.165}$$

$$\frac{1}{\rho}\left[\frac{\partial}{\partial \rho}(\rho E_\phi) - \frac{\partial E_\rho}{\partial \phi}\right] = -i\omega\mu H_z \tag{4.166}$$

$$\frac{1}{\rho}\frac{\partial E_z}{\partial \phi} = -i\omega\mu H_\rho \tag{4.167}$$

$$\frac{\partial E_z}{\partial \rho} = i\omega\mu H_\phi \tag{4.168}$$

$$\frac{1}{\rho}\left[\frac{\partial}{\partial \rho}(\rho H_\phi) - \frac{\partial H_\rho}{\partial \phi}\right] = J_z + i\omega\epsilon E_z \tag{4.169}$$

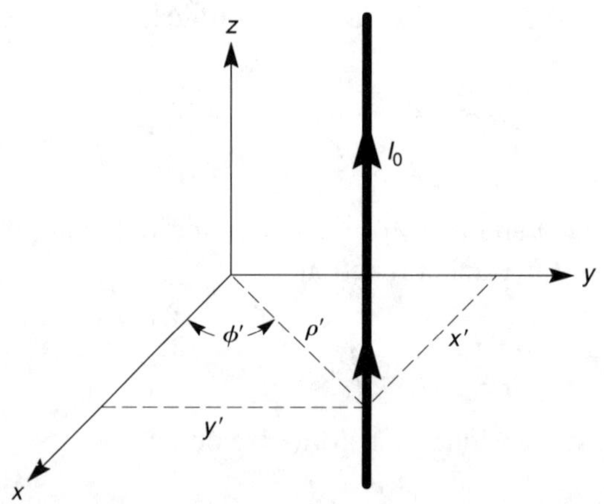

Fig. 4-6 Electric line current located at (x', y') and extending from $z = -\infty$ to $z = \infty$.

where we have used $\partial/\partial z = 0$ in making the expansions. We have arranged the six equations above such that the first three are TE_z and the second three are TM_z. We note that the TE_z and TM_z fields are not coupled, and that the TE_z fields are source-free. We therefore conclude that the TE_z fields are zero. For the TM_z set, we multiply (4.168) by ρ and differentiate with respect to ρ to give

$$\frac{\partial}{\partial \rho}\left(\rho \frac{\partial E_z}{\partial \rho}\right) = i\omega\mu \frac{\partial}{\partial \rho}(\rho H_\phi) \tag{4.170}$$

Next, we differentiate (4.167) with respect to ϕ to give

$$\frac{1}{\rho}\frac{\partial^2 E_z}{\partial \phi^2} = -i\omega\mu \frac{\partial H_\rho}{\partial \phi} \tag{4.171}$$

Using (4.170) and (4.171) in (4.169) to eliminate H_ϕ and H_ρ yields the following:

$$\nabla^2_{\rho\phi} E_z + k^2 E_z = i\omega\mu I_0 \frac{\delta(\rho - \rho')\delta(\phi - \phi')}{\rho} \tag{4.172}$$

where, from (4.167) and (4.168), we have

$$H_\phi = \frac{1}{i\omega\mu}\frac{\partial E_z}{\partial \rho} \tag{4.173}$$

$$H_\rho = -\frac{1}{i\omega\mu\rho}\frac{\partial E_z}{\partial \phi} \tag{4.174}$$

and where

$$\nabla^2_{\rho\phi} = \frac{1}{\rho}\left[\frac{\partial}{\partial \rho}\left(\rho \frac{\partial}{\partial \rho}\right)\right] + \frac{1}{\rho^2}\frac{\partial^2}{\partial \phi^2} \tag{4.175}$$

The boundary and limiting conditions associated with E_z are as follows:

$$\lim_{\rho \to \infty} E_z = 0 \tag{4.176}$$

$$\lim_{\rho \to 0} E_z = \text{finite} \tag{4.177}$$

$$E_z \bigg|_{\phi=\phi_0} = E_z \bigg|_{\phi=\phi_0+2\pi} \tag{4.178}$$

$$\frac{\partial E_z}{\partial \phi}\bigg|_{\phi=\phi_0} = \frac{\partial E_z}{\partial \phi}\bigg|_{\phi=\phi_0+2\pi} \tag{4.179}$$

where ϕ_0 is any fixed angle. Let

$$g = -\frac{E_z}{i\omega\mu I_0} \tag{4.180}$$

Substitution of (4.180) and (4.175) into (4.172) yields the two-dimensional Green's function problem

$$\frac{1}{\rho}\left[\frac{\partial}{\partial\rho}\left(\rho\frac{\partial g}{\partial\rho}\right)\right] + \frac{1}{\rho^2}\frac{\partial^2 g}{\partial\phi^2} + k^2 g = -\frac{\delta(\rho-\rho')}{\rho}\delta(\phi-\phi') \tag{4.181}$$

$$\lim_{\rho\to\infty} g = 0 \tag{4.182}$$

$$\lim_{\rho\to 0} g = \text{finite} \tag{4.183}$$

$$g\Big|_{\phi=\phi_0} = g\Big|_{\phi=\phi_0+2\pi} \tag{4.184}$$

$$\frac{\partial g}{\partial\phi}\bigg|_{\phi=\phi_0} = \frac{\partial g}{\partial\phi}\bigg|_{\phi=\phi_0+2\pi} \tag{4.185}$$

In order to separate the ρ-operator from the ϕ-operator, we multiply both sides of (4.181) by ρ^2 and obtain

$$\rho\left[\frac{\partial}{\partial\rho}\left(\rho\frac{\partial g}{\partial\rho}\right)\right] + \frac{\partial^2 g}{\partial\phi^2} + (k\rho)^2 g = -\rho\delta(\rho-\rho')\delta(\phi-\phi') \tag{4.186}$$

A perhaps more descriptive way of writing (4.186) is as follows:

$$(L_\rho + L_\phi)g = \rho\delta(\rho-\rho')\delta(\phi-\phi') \tag{4.187}$$

where

$$L_\rho = -\rho\left[\frac{\partial}{\partial\rho}\left(\rho\frac{\partial}{\partial\rho}\right)\right] - (k\rho)^2 \tag{4.188}$$

$$L_\phi = -\frac{\partial^2}{\partial\phi^2} \tag{4.189}$$

The operator L_ρ, with boundary and limiting conditions given in (4.182) and (4.183), is called the *Kantorovich–Lebedev* operator and leads to the Kantorovich–Lebedev transform considered in Example 3.6. The operator L_ϕ with periodic boundary conditions given in (4.184) and (4.185) produces the complex Fourier series considered in Problem 3.2. To solve for the Green's function $g(\rho, \phi, \rho', \phi')$, we have the choice of applying a complex

Fourier series expansion or the Kantorovich–Lebedev transform. We shall consider both choices in turn. We begin by choosing the complex Fourier series.

Using the complex Fourier expansion in Problem 3.2, we expand the Green's function as follows:

$$g(\rho, \phi, \rho', \phi') = \sum_{n=-\infty}^{\infty} a_n(\rho, \rho', \phi') \sqrt{\frac{1}{2\pi}} e^{in\phi} \tag{4.190}$$

The coefficient a_n is given by

$$\begin{aligned} a_n(\rho, \rho', \phi') &= \int_0^{2\pi} g(\rho, \phi, \rho', \phi') \sqrt{\frac{1}{2\pi}} e^{-in\phi} d\phi \\ &= \langle g, u_n \rangle \end{aligned} \tag{4.191}$$

where u_n is the normalized eigenfunction

$$u_n = \sqrt{\frac{1}{2\pi}} e^{in\phi} \tag{4.192}$$

and where we have defined the complex inner product

$$\langle u, v \rangle = \int_0^{2\pi} u(\phi) \overline{v}(\phi) d\phi \tag{4.193}$$

We symbolize the transform from g to a_n given in (4.191) by

$$g \Longrightarrow a_n \tag{4.194}$$

Since the operator L_ϕ is self-adjoint, we use the procedure in (3.24)–(3.27) and find that

$$L_\phi g \Longrightarrow n^2 a_n \tag{4.195}$$

Also,

$$\delta(\phi - \phi') \Longrightarrow \sqrt{\frac{1}{2\pi}} e^{-in\phi'} \tag{4.196}$$

Using these results to transform (4.187), we obtain

$$(L_\rho + n^2) a_n = \rho \delta(\rho - \rho') \sqrt{\frac{1}{2\pi}} e^{-in\phi'} \tag{4.197}$$

Substituting (4.188) and dividing both sides by ρ^2 gives

$$-\left\{\frac{1}{\rho}\left[\frac{\partial}{\partial\rho}\left(\rho\frac{\partial}{\partial\rho}\right)\right]+k^2-\frac{n^2}{\rho^2}\right\}b_n=\frac{\delta(\rho-\rho')}{\rho} \tag{4.198}$$

where

$$b_n=\frac{a_n}{\sqrt{\frac{1}{2\pi}}e^{-in\phi'}} \tag{4.199}$$

For limiting conditions associated with b_n, we choose

$$\lim_{\rho\to\infty} b_n=0 \tag{4.200}$$

$$\lim_{\rho\to 0} b_n=\text{ finite} \tag{4.201}$$

The reader should verify that this choice of limiting conditions is consistent with the limiting conditions associated with g given in (4.182) and (4.183).

We now solve (4.198) by the Green's function method. This particular Green's function problem has been previously considered in Example 3.6. Using the results therein, we may immediately write

$$b_n=\frac{\pi}{2i}\begin{cases} H_n^{(2)}(k\rho')J_n(k\rho), & \rho<\rho' \\ H_n^{(2)}(k\rho)J_n(k\rho'), & \rho>\rho' \end{cases} \tag{4.202}$$

Substituting (4.202) into (4.199), solving for a_n, and substituting the result into (4.190) yields

$$g(\rho,\phi,\rho',\phi')=\frac{1}{4i}\sum_{n=-\infty}^{\infty}e^{in(\phi-\phi')}\begin{cases} H_n^{(2)}(k\rho')J_n(k\rho), & \rho<\rho' \\ H_n^{(2)}(k\rho)J_n(k\rho'), & \rho>\rho' \end{cases} \tag{4.203}$$

Finally, this result substituted into (4.180) gives

$$E_z=-\frac{\omega\mu I_0}{4}\sum_{n=-\infty}^{\infty}e^{in(\phi-\phi')}\begin{cases} H_n^{(2)}(k\rho')J_n(k\rho), & \rho<\rho' \\ H_n^{(2)}(k\rho)J_n(k\rho'), & \rho>\rho' \end{cases} \tag{4.204}$$

We recall from (4.117) that the electric field from a line source located at the origin is given by

$$E_z=-\frac{\omega\mu I_0}{4}H_0^{(2)}(k\rho)$$

A coordinate transformation $x \to x - x'$, $y \to y - y'$ gives, in cylindrical coordinates,

$$E_z = -\frac{\omega\mu I_0}{4} H_0^{(2)}(k(|\boldsymbol{\rho} - \boldsymbol{\rho}'|) \tag{4.205}$$

where

$$|\boldsymbol{\rho} - \boldsymbol{\rho}'| = \sqrt{(x - x')^2 + (y - y')^2} = \sqrt{\rho^2 + \rho'^2 - 2\rho\rho' \cos(\phi - \phi')} \tag{4.206}$$

Since (4.204) and (4.205) must yield the same result, we may equate them to give

$$H_0^{(2)}(k|\boldsymbol{\rho} - \boldsymbol{\rho}'|) = \sum_{n=-\infty}^{\infty} e^{in(\phi-\phi')} \begin{cases} H_n^{(2)}(k\rho')J_n(k\rho), & \rho < \rho' \\ H_n^{(2)}(k\rho)J_n(k\rho'), & \rho > \rho' \end{cases} \tag{4.207}$$

which is the *Addition Theorem* for the Hankel function.

We next obtain an alternative representation of the solution for the electric field by applying the Kantorovich–Lebedev transform to (4.186). If $g(\rho, \phi, \rho', \phi')$ and $G(\beta, \phi, \rho', \phi')$ are Kantorovich–Lebedev transform pairs, then the transform applied to (4.186) gives

$$\frac{d^2G}{d\phi^2} + \beta^2 G = -H_\beta^{(2)}(k\rho')\delta(\phi - \phi') \tag{4.208}$$

where we have used the transform in (3.157) and the derivative transform in (3.161). We define

$$\hat{G}(\beta, \phi, \phi') = \frac{G(\beta, \phi, \rho', \phi')}{H_\beta^{(2)}(k\rho')} \tag{4.209}$$

and obtain

$$\frac{d^2\hat{G}}{d\phi^2} + \beta^2 \hat{G} = -\delta(\phi - \phi') \tag{4.210}$$

with the boundary conditions

$$\hat{G}(\beta, \phi_0, \phi') = \hat{G}(\beta, \phi_0 + 2\pi, \phi') \tag{4.211}$$

$$\frac{d\hat{G}(\beta, \phi_0, \phi')}{d\phi} = \frac{d\hat{G}(\beta, \phi_0 + 2\pi, \phi')}{d\phi} \tag{4.212}$$

From Problem 2.18, the solution to this Green's function problem is given by

$$\hat{G} = -\frac{\cos[\beta(|\phi - \phi'| - \pi)]}{2\beta \sin \pi\beta} \tag{4.213}$$

We substitute (4.213) into (4.209), solve for G, and use the inverse Kantorovich–Lebedev transform given in (3.159) to obtain

$$g(\rho, \phi, \rho', \phi') = -\frac{1}{8} \int_{i\infty}^{-i\infty} \frac{H_\beta^{(2)}(k\rho) H_\beta^{(2)}(k\rho') \cos[\beta(|\phi - \phi'| - \pi)]}{\sin \pi\beta} d\beta \tag{4.214}$$

Substituting into (4.180) and solving for E_z gives the electric field at (ρ, ϕ) caused by an electric line current source at (ρ', ϕ'), viz.

$$E_z = \frac{i\omega\mu I_0}{8} \int_{i\infty}^{-i\infty} \frac{H_\beta^{(2)}(k\rho) H_\beta^{(2)}(k\rho') \cos[\beta(|\phi - \phi'| - \pi)]}{\sin \pi\beta} d\beta \tag{4.215}$$

Equation (4.215) gives an alternative representation to (4.204) for the electric field. By comparison with (4.205), we obtain the following integral representation alternative to the Hankel function addition theorem in (4.207):

$$H_0^{(2)}(k|\boldsymbol{\rho} - \boldsymbol{\rho}'|) = \frac{1}{2i} \int_{i\infty}^{-i\infty} \frac{H_\beta^{(2)}(k\rho) H_\beta^{(2)}(k\rho') \cos[\beta(|\phi - \phi'| - \pi)]}{\sin \pi\beta} d\beta \tag{4.216}$$

For further discussion of the Kantorovich–Lebedev transform, the reader is referred to [9]. The transform is particularly useful in the solution to electromagnetic problems involving conducting wedges [10].

4.7 THE CYLINDRICAL SHELL SOURCE

Consider a circularly-cylindrical shell current source (Fig. 4-7), located symmetrically about the z-axis and extending over $z \in (-\infty, \infty)$. We represent the current by

$$\mathbf{J} = \hat{z} I_0 \frac{\delta(\rho - \rho')}{2\pi\rho} \tag{4.217}$$

The factor 2π has been included so that the total current is I_0, viz.

$$\int_0^{2\pi} \int_0^\infty I_0 \left[\frac{\delta(\rho - \rho')}{2\pi\rho}\right] \rho d\rho d\phi = I_0$$

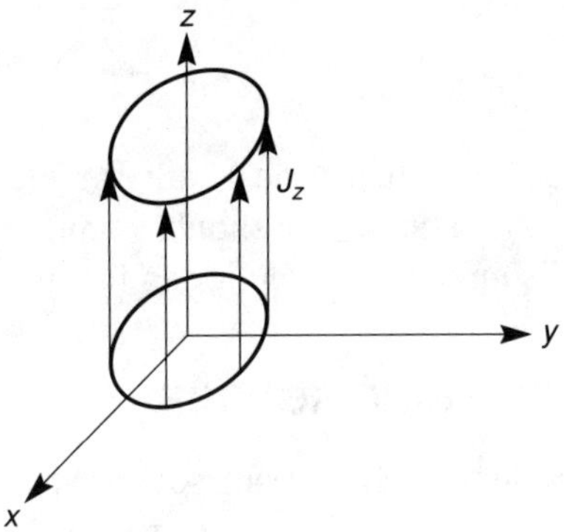

Fig. 4-7 Electric cylindrical shell current located symmetrically about the z-axis and extending from $z = -\infty$ to $z = \infty$.

This problem is independent of both ϕ and z. Therefore, using (4.109) and (4.110), we have

$$\frac{1}{\rho}\left[\frac{d}{d\rho}\left(\rho\frac{dE_z}{d\rho}\right)\right] + k^2 E_z = i\omega\mu I_0 \frac{\delta(\rho-\rho')}{2\pi\rho} \tag{4.218}$$

$$H_\phi = \frac{1}{i\omega\mu}\frac{dE_z}{d\rho} \tag{4.219}$$

Let

$$g = -\frac{2\pi E_z}{i\omega\mu I_0} \tag{4.220}$$

so that

$$\frac{1}{\rho}\left[\frac{d}{d\rho}\left(\rho\frac{dg}{d\rho}\right)\right] + k^2 g = -\frac{\delta(\rho-\rho')}{\rho} \tag{4.221}$$

with the limiting condition

$$\lim_{\rho\to\infty} g(\rho,\rho') = 0$$

and a finiteness condition at the origin. The solution to this Green's function problem has been given in (4.115), viz.

$$g(\rho,\rho') = \frac{\pi}{2i}\begin{cases} H_0^{(2)}(k\rho')J_0(k\rho), & \rho < \rho' \\ H_0^{(2)}(k\rho)J_0(k\rho'), & \rho > \rho' \end{cases} \tag{4.222}$$

Substituting into (4.220) and solving for E_z, we obtain

$$E_z = -\frac{\omega\mu I_0}{4}\begin{cases} H_0^{(2)}(k\rho')J_0(k\rho), & \rho < \rho' \\ H_0^{(2)}(k\rho)J_0(k\rho'), & \rho > \rho' \end{cases} \tag{4.223}$$

Note that

$$\lim_{\rho' \to 0} E_z = -\frac{\omega \mu I_0}{4} H_0^{(2)}(k\rho) \tag{4.224}$$

which reproduces the result for the line current at the origin, given in (4.117). The reader should compare the development of the cylindrical shell source in this section to the treatment in [11].

4.8 THE RING SOURCE

Consider a magnetic ring source **M**, located symmetrically about the z-axis (Fig. 4-8). We describe the source by

$$\mathbf{M} = P_0 \frac{\delta(\rho - \rho')}{\rho} \delta(z - z')\hat{\phi} \tag{4.225}$$

where $\hat{\phi}$ is a unit vector in the ϕ-direction and P_0 is a magnetic current moment in volt · meters. Since the problem is symmetric in ϕ, we must have $\partial/\partial\phi = 0$. With this restriction, Maxwell's curl equations in cylindrical coordinates reduce to

$$-\frac{\partial H_\phi}{\partial z} = i\omega\epsilon E_\rho \tag{4.226}$$

$$\frac{1}{\rho}\frac{\partial}{\partial \rho}(\rho H_\phi) = i\omega\epsilon E_z \tag{4.227}$$

$$\frac{\partial E_\rho}{\partial z} - \frac{\partial E_z}{\partial \rho} = -M_\phi - i\omega\mu H_\phi \tag{4.228}$$

$$\frac{\partial H_\rho}{\partial z} - \frac{\partial H_z}{\partial \rho} = i\omega\epsilon E_\phi \tag{4.229}$$

$$\frac{\partial E_\phi}{\partial z} = i\omega\mu H_\rho \tag{4.230}$$

$$\frac{1}{\rho}\frac{\partial}{\partial \rho}(\rho E_\phi) = -i\omega\mu H_z \tag{4.231}$$

We have arranged the six equations above so that the first three are TE_ϕ and the second three are TM_ϕ. We note that the TE_ϕ and TM_ϕ fields are not coupled, and that the TM_ϕ fields are source-free. We therefore conclude that the TM_ϕ fields are zero. For the TE_ϕ set, we differentiate (4.226) with respect to z and (4.227) with respect to ρ to obtain

$$-\frac{\partial^2 H_\phi}{\partial z^2} = i\omega\epsilon \frac{\partial E_\rho}{\partial z} \tag{4.232}$$

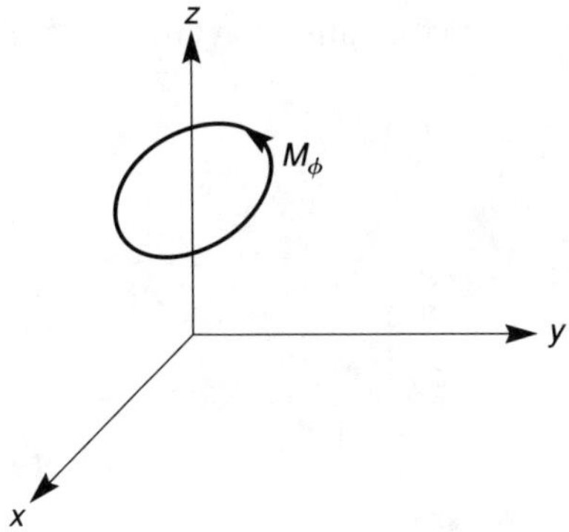

Fig. 4-8 Magnetic ring current source located symmetrically about the z-axis.

$$\frac{\partial}{\partial\rho}\left[\frac{1}{\rho}\frac{\partial}{\partial\rho}(\rho H_\phi)\right] = i\omega\epsilon\frac{\partial E_z}{\partial\rho} \tag{4.233}$$

Subtracting (4.232) from (4.233) and substituting (4.228), we obtain

$$\frac{\partial}{\partial\rho}\left[\frac{1}{\rho}\frac{\partial}{\partial\rho}(\rho H_\phi)\right] + \frac{\partial^2 H_\phi}{\partial z^2} + k^2 H_\phi = i\omega\epsilon M_\phi \tag{4.234}$$

But,

$$\frac{\partial}{\partial\rho}\left[\frac{1}{\rho}\frac{\partial}{\partial\rho}(\rho H_\phi)\right] = \frac{1}{\rho}\frac{\partial}{\partial\rho}\left(\rho\frac{\partial H_\phi}{\partial\rho}\right) - \frac{H_\phi}{\rho^2} \tag{4.235}$$

Substituting (4.235) into (4.234), we obtain

$$\nabla^2_{\rho z} H_\phi + \left(k^2 - \frac{1}{\rho^2}\right) H_\phi = i\omega\epsilon M_\phi \tag{4.236}$$

where

$$\nabla^2_{\rho z} = \frac{1}{\rho}\frac{\partial}{\partial\rho}\left(\rho\frac{\partial}{\partial\rho}\right) + \frac{\partial^2}{\partial z^2} \tag{4.237}$$

and

$$M_\phi = P_0\frac{\delta(\rho - \rho')}{\rho}\delta(z - z') \tag{4.238}$$

Once we have solved (4.236), the electric field components can be obtained from (4.226) and (4.227), viz.

$$E_\rho = -\frac{1}{i\omega\epsilon}\frac{\partial H_\phi}{\partial z} \tag{4.239}$$

$$E_z = \frac{1}{i\omega\epsilon\rho}\frac{\partial}{\partial\rho}(\rho H_\phi) \tag{4.240}$$

By suitably normalizing H_ϕ, we may obtain the following Green's function problem:

$$\left(L_\rho + L_z - k^2\right) g = \frac{\delta(\rho - \rho')}{\rho} \delta(z - z') \tag{4.241}$$

where

$$L_\rho = -\frac{1}{\rho}\frac{\partial}{\partial \rho}\left(\rho \frac{\partial}{\partial \rho}\right) + \frac{1}{\rho^2} \tag{4.242}$$

$$L_z = -\frac{\partial^2}{\partial z^2} \tag{4.243}$$

and where

$$g = -\frac{H_\phi}{i\omega\epsilon P_0} \tag{4.244}$$

The associated limiting conditions are

$$\lim_{\rho \to 0} g \text{ finite} \tag{4.245}$$

$$\lim_{\rho \to \infty} g = 0 \tag{4.246}$$

$$\lim_{z \to \pm\infty} g = 0 \tag{4.247}$$

As shown in Problem 3.5, the operator L_ρ in (4.242) with limiting conditions in (4.245) and (4.246) leads to the Fourier–Bessel transform of order one. As shown in Example 3.4, the operator L_z in (4.243) with limiting conditions in (4.247) leads to the Fourier transform. We therefore have the choice of applying either of these transforms. We shall consider each in turn, beginning with the Fourier transform.

Using the Fourier transform over the z-variable in (4.241), we obtain

$$-\frac{1}{\rho}\frac{\partial}{\partial \rho}\left(\rho \frac{\partial G}{\partial \rho}\right) - \left(k_\rho^2 - \frac{1}{\rho^2}\right) G = e^{-ik_z z'} \frac{\delta(\rho - \rho')}{\rho} \tag{4.248}$$

where $g(\rho, z, \rho', z')$ and $G(\rho, k_z, \rho', z')$ are Fourier transform pairs and

$$k_\rho^2 = k^2 - k_z^2 \tag{4.249}$$

We let

$$H = \frac{G}{e^{-ik_z z'}} \tag{4.250}$$

and obtain

$$-\frac{1}{\rho}\frac{\partial}{\partial \rho}\left(\rho \frac{\partial H}{\partial \rho}\right) - \left(k_\rho^2 - \frac{1}{\rho^2}\right) H = \frac{\delta(\rho - \rho')}{\rho} \tag{4.251}$$

We may satisfy the limiting conditions in (4.245) and (4.246) by requiring

$$\lim_{\rho\to 0} H \text{ finite} \tag{4.252}$$

$$\lim_{\rho\to\infty} H = 0 \tag{4.253}$$

The solution to (4.251) with conditions in (4.252) and (4.253) has been previously considered in Problem 3.5. We find that

$$H = \frac{\pi}{2i}\begin{cases} H_1^{(2)}(k_\rho \rho')J_1(k_\rho \rho), & \rho < \rho' \\ H_1^{(2)}(k_\rho \rho)J_1(k_\rho \rho'), & \rho > \rho' \end{cases} \tag{4.254}$$

Substituting (4.254) into (4.250) and the result into (4.244), we obtain

$$H_\phi = -\frac{\omega\epsilon P_0}{4}\int_{-\infty}^{\infty} dk_z e^{ik_z(z-z')}\begin{cases} H_1^{(2)}(k_\rho \rho')J_1(k_\rho \rho), & \rho < \rho' \\ H_1^{(2)}(k_\rho \rho)J_1(k_\rho \rho'), & \rho > \rho' \end{cases} \tag{4.255}$$

We may obtain an alternative representation by taking the Fourier–Bessel transform of order one in (4.241). From the results in Problem 3.5, we may write the following Fourier–Bessel transform pair:

$$\hat{G}(\lambda, z, \rho', z') = \int_0^\infty g(\rho, z, \rho', z')J_1(\lambda\rho)\rho d\rho \tag{4.256}$$

with inverse

$$g(\rho, z, \rho', z') = \int_0^\infty \hat{G}(\lambda, z, \rho', z')J_1(\lambda\rho)\lambda d\lambda \tag{4.257}$$

From Problem 3.5, we have the following Fourier–Bessel transform pair:

$$\left[-\frac{1}{\rho}\frac{d}{d\rho}\left(\rho\frac{d}{d\rho}\right) + \frac{1}{\rho^2}\right] g \Longleftrightarrow \lambda^2 \hat{G} \tag{4.258}$$

Applying (4.256) and (4.258) in (4.241), we obtain

$$-\frac{d^2\hat{H}}{dz^2} - \beta^2\hat{H} = \delta(z - z') \tag{4.259}$$

where

$$\hat{H} = \frac{\hat{G}}{J_1(\lambda\rho')} \tag{4.260}$$

and

$$\beta = \sqrt{k^2 - \lambda^2} \tag{4.261}$$

To satisfy the limiting condition in (4.247), we require

$$\lim_{z\to\pm\infty} \hat{H} = 0 \tag{4.262}$$

We have previously considered this Green's function problem in Example 2.20. The result applied to (4.259) is as follows:

$$\hat{H} = \frac{e^{-i\beta|z-z'|}}{2i\beta}, \qquad \text{Im}(\beta) < 0 \tag{4.263}$$

Substituting (4.263) into (4.260) and the result into (4.257), we obtain

$$g = \int_0^\infty \frac{e^{-i\beta|z-z'|}}{2i\beta} J_1(\lambda\rho')J_1(\lambda\rho)\lambda d\lambda \tag{4.264}$$

Finally, the magnetic field can be obtained by substituting (4.264) into (4.244), viz.

$$H_\phi = -i\omega\epsilon P_0 \int_0^\infty \frac{e^{-i\beta|z-z'|}}{2i\beta} J_1(\lambda\rho')J_1(\lambda\rho)\lambda d\lambda \tag{4.265}$$

4.9 THE POINT SOURCE

Consider an electric point source located at the origin (Fig. 4-9) and polarized in the z-direction. We assume that the source radiates in empty space and that there is a small amount of loss present so that we may apply limiting conditions on the fields as we approach $\pm\infty$ in x, y, and z. We may describe the source in terms of its current moment $I_0\ell$ as follows:

$$\mathbf{J} = \hat{z}I_0\ell\delta(\mathbf{r}) \tag{4.266}$$

where the units of $I_0\ell$ are amp · meters and where

$$\delta(\mathbf{r}) = \delta(x)\delta(y)\delta(z) \tag{4.267}$$

The fields from such a source can be described conveniently by use of the magnetic vector potential **A**, as follows:

$$-\left(\nabla^2 + k^2\right)\mathbf{A} = \mu\mathbf{J} \tag{4.268}$$

The magnetic and electric fields are obtained from the vector potential by the following two relationships:

$$\mathbf{H} = \frac{1}{\mu}\nabla\times\mathbf{A} \tag{4.269}$$

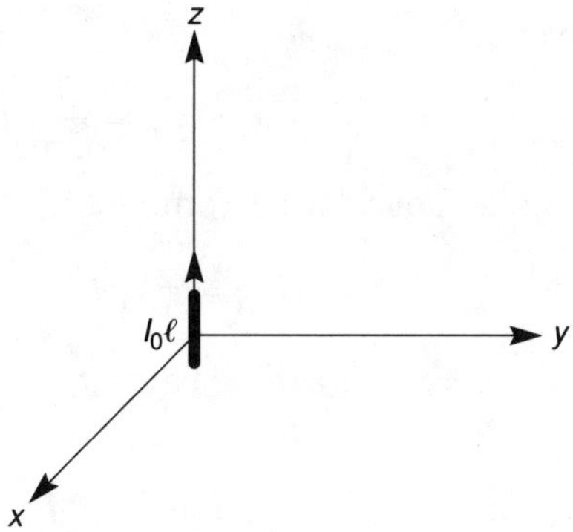

Fig. 4-9 Electric current point source located at the origin.

$$\mathbf{E} = -i\omega\left[\mathbf{A} + \frac{1}{k^2}\nabla(\nabla\cdot\mathbf{A})\right] \tag{4.270}$$

Since the current is in the z-direction, the magnetic vector potential is also z-directed, viz.

$$\mathbf{A} = \hat{z}A_z \tag{4.271}$$

The z-component of (4.268) yields a differential equation for determining the vector potential A_z, viz.

$$-\left(\nabla^2 + k^2\right)A_z = \mu I_0\ell\delta(\mathbf{r}) \tag{4.272}$$

We let

$$g = \frac{A_z}{\mu I_0\ell} \tag{4.273}$$

and obtain

$$-\left(\nabla^2 + k^2\right)g = \delta(\mathbf{r}) \tag{4.274}$$

We first consider the solution to the point source problem in Cartesian coordinates. Expanding (4.274), we have

$$-\left(\frac{\partial^2}{\partial x^2} + \frac{\partial^2}{\partial y^2} + \frac{\partial^2}{\partial z^2} + k^2\right)g = \delta(x)\delta(y)\delta(z) \tag{4.275}$$

From the results in Exampe 3.4, the spectral representation of each of the three differential operators in (4.275), with limiting conditions at $\pm\infty$, yields the Fourier transform. Typically, for the operator $-\partial^2/\partial x^2$ with limiting conditions

$$\lim_{x\to\pm\infty} g = 0$$

we have

$$\delta(x - x') = \frac{1}{2\pi} \int_{-\infty}^{\infty} e^{ik_x(x-x')} dk_x \tag{4.276}$$

Taking the Fourier transform in (4.275) over x, y, z, we obtain

$$-\left(-k_x^2 - k_y^2 - k_z^2 + k^2\right) G = 1$$

where $g \Longleftrightarrow G$ is a triple Fourier transform pair over (x, y, z), (k_x, k_y, k_z). Solving for G, we obtain

$$G = -\frac{1}{k^2 - k_x^2 - k_y^2 - k_z^2}$$

Taking the inverse transform over k_x, k_y, k_z yields

$$g = -\frac{1}{(2\pi)^3} \int_{-\infty}^{\infty} \int_{-\infty}^{\infty} \int_{-\infty}^{\infty} \frac{e^{i(k_x x + k_y y + k_z z)}}{k^2 - k_x^2 - k_y^2 - k_z^2} dk_x dk_y dk_z \tag{4.277}$$

Using (4.273), we obtain, for the magnetic vector potential,

$$A_z = -\frac{\mu I_0 \ell}{(2\pi)^3} \int_{-\infty}^{\infty} \int_{-\infty}^{\infty} \int_{-\infty}^{\infty} \frac{e^{i(k_x x + k_y y + k_z z)}}{k^2 - k_x^2 - k_y^2 - k_z^2} dk_x dk_y dk_z \tag{4.278}$$

An alternative form in Cartesian coordinates can be obtained by taking the Fourier transform over any two Cartesian variables. In (4.275), if we take the Fourier transform over x and y, we obtain

$$-\left(\frac{d^2}{dz^2} + k^2 - k_x^2 - k_y^2\right) \hat{G} = \delta(z) \tag{4.279}$$

where $g \Longleftrightarrow \hat{G}$ is a double Fourier transform pair over (x, y), (k_x, k_y). The one-dimensional Green's function problem in (4.279) is identical to that posed in (4.152) and (4.153). The solution is given by

$$\hat{G} = \frac{e^{-i\beta|z|}}{2i\beta}, \qquad \text{Im}(\beta) < 0 \tag{4.280}$$

where

$$\beta = \sqrt{k^2 - k_x^2 - k_y^2} \tag{4.281}$$

Taking the inverse Fourier transform over k_x, k_y yields

$$g = \frac{1}{(2\pi)^2} \int_{-\infty}^{\infty} \int_{-\infty}^{\infty} \frac{e^{-i\beta|z|}}{2i\beta} e^{i(k_x x + k_y y)} dk_x dk_y \tag{4.282}$$

Using (4.273), we obtain the magnetic vector potential A_z as follows:

$$A_z = \frac{\mu I_0 \ell}{(2\pi)^2} \int_{-\infty}^{\infty} \int_{-\infty}^{\infty} \frac{e^{-i\beta|z|}}{2i\beta} e^{i(k_x x + k_y y)} dk_x dk_y \tag{4.283}$$

We next consider the point source problem in cylindrical coordinates. We note that, in cylindrical coordinates, the Green's function problem in (4.274) is a function of ρ and z and independent of ϕ. Expanding (4.274), we obtain

$$-\frac{1}{\rho}\frac{\partial}{\partial \rho}\left(\rho \frac{\partial g}{\partial \rho}\right) - \frac{\partial^2 g}{\partial z^2} - k^2 g = \frac{\delta(\rho)\delta(z)}{2\pi\rho} \tag{4.284}$$

The spectral representation of the ρ-operator produces the Fourier–Bessel transform of order zero, while the spectral representation of the z-operator produces the Fourier transform. We shall derive three forms of solution in cylindrical coordinates. For the first form, we apply the Fourier–Bessel transform to (4.284) and obtain

$$-\frac{d^2 \tilde{G}}{dz^2} - \Gamma^2 \tilde{G} = \frac{\delta(z)}{2\pi} \tag{4.285}$$

where

$$\tilde{G}(\lambda, z) = \int_0^{\infty} g(\rho, z) J_0(\lambda\rho) \rho d\rho \tag{4.286}$$

and

$$\Gamma = \sqrt{k^2 - \lambda^2} \tag{4.287}$$

The solution to (4.285) with limiting conditions

$$\lim_{z \to \pm\infty} \tilde{G} = 0$$

is the same as (4.263), viz.

$$2\pi \tilde{G} = \frac{e^{-i\Gamma|z|}}{2i\Gamma}, \qquad \mathrm{Im}(\Gamma) < 0 \tag{4.288}$$

Taking the inverse Fourier–Bessel transform yields

$$g(\rho, z) = \frac{1}{2\pi} \int_0^{\infty} \frac{e^{-i\Gamma|z|}}{2i\Gamma} J_0(\lambda\rho) \lambda d\lambda \tag{4.289}$$

Using (4.273), we produce the magnetic vector potential

$$A_z = \frac{\mu I_0 \ell}{2\pi} \int_0^{\infty} \frac{e^{-i\Gamma|z|}}{2i\Gamma} J_0(\lambda\rho) \lambda d\lambda \tag{4.290}$$

The second form of solution is obtained by taking the Fourier transform of (4.284) with respect to z, with the result

$$-\frac{1}{\rho}\frac{\partial}{\partial\rho}\left(\rho\frac{\partial\hat{G}}{\partial\rho}\right)-\tau^2\hat{G}=\frac{\delta(\rho)}{2\pi\rho} \tag{4.291}$$

where

$$\hat{G}(\rho,k_z)=\int_{-\infty}^{\infty}g(\rho,z)e^{-ik_zz}dz \tag{4.292}$$

and

$$\tau=\sqrt{k^2-k_z^2} \tag{4.293}$$

We now use (2.185) and write the solution to (4.291) as follows:

$$2\pi\hat{G}=\frac{\pi}{2i}H_0^{(2)}(\tau\rho) \tag{4.294}$$

Taking the inverse Fourier transform, we obtain

$$g=\frac{1}{8\pi i}\int_{-\infty}^{\infty}H_0^{(2)}(\tau\rho)e^{ik_zz}dk_z \tag{4.295}$$

Using (4.273), we have, for the magnetic vector potential,

$$A_z=\frac{\mu I_0\ell}{8\pi i}\int_{-\infty}^{\infty}H_0^{(2)}(\tau\rho)e^{ik_zz}dk_z \tag{4.296}$$

The third form of solution is obtained by taking the Fourier–Bessel transform with respect to ρ *and* the Fourier transform with respect to z, with the result

$$(\lambda^2+k_z^2-k^2)\mathcal{G}=\frac{1}{2\pi}$$

Solving for $\mathcal{G}$ and taking the inverse Fourier–Bessel transform and the inverse Fourier transform gives

$$g=\frac{1}{4\pi^2}\int_{-\infty}^{\infty}\int_0^{\infty}\frac{e^{ik_zz}J_0(\lambda\rho)}{\lambda^2+k_z^2-k^2}\lambda d\lambda dk_z \tag{4.297}$$

Finally, we consider the point source problem in spherical coordinates. We note that the Green's function problem posed by (4.274) is symmetric in spherical coordinates over both θ and ϕ. Therefore, the operator ∇^2 is given totally by its radial component. We therefore may write (4.274) as follows:

$$\frac{1}{r^2}\left[\frac{d}{dr}\left(r^2\frac{dg}{dr}\right)\right]+k^2g=-\frac{\delta(r)}{4\pi r^2} \tag{4.298}$$

where we have used (4.22) for the spherical coordinate representation of the delta function at the origin. We have obtained the solution to this Green's function problem previously in Example 2.22. Using (2.201), we obtain

$$g = \frac{e^{-ikr}}{4\pi r} \tag{4.299}$$

Again using (4.273), we obtain the magnetic vector potential, as follows:

$$A_z = \frac{\mu I_0 \ell}{4\pi}\left(\frac{e^{-ikr}}{r}\right) \tag{4.300}$$

We may exhibit five identities from the alternative representations of the point source. Comparing (4.299) and (4.277), we obtain

$$\frac{e^{-ikr}}{r} = -\frac{1}{2\pi^2}\int_{-\infty}^{\infty}\int_{-\infty}^{\infty}\int_{-\infty}^{\infty}\frac{e^{i(k_x x+k_y y+k_z z)}}{k^2-k_x^2-k_y^2-k_z^2}dk_x dk_y dk_z \tag{4.301}$$

From (4.299) and (4.282), we have

$$\frac{e^{-ikr}}{r} = \frac{1}{\pi}\int_{-\infty}^{\infty}\int_{-\infty}^{\infty}\frac{e^{-i\beta|z|}}{2i\beta}e^{i(k_x x+k_y y)}dk_x dk_y \tag{4.302}$$

where

$$\beta = \sqrt{k^2-k_x^2-k_y^2} \tag{4.303}$$

From (4.299) and (4.289), we have

$$\frac{e^{-ikr}}{r} = 2\int_0^{\infty}\frac{e^{-i\Gamma|z|}}{2i\Gamma}J_0(\lambda\rho)\lambda d\lambda \tag{4.304}$$

where

$$\Gamma = \sqrt{k^2-\lambda^2} \tag{4.305}$$

From (4.299) and (4.295), we have

$$\frac{e^{-ikr}}{r} = \frac{1}{2i}\int_{-\infty}^{\infty}H_0^{(2)}(\tau\rho)e^{ik_z z}dk_z \tag{4.306}$$

where

$$\tau = \sqrt{k^2-k_z^2} \tag{4.307}$$

From (4.299) and (4.297), we have

$$\frac{e^{-ikr}}{r} = \frac{1}{\pi}\int_{-\infty}^{\infty}\int_0^{\infty}\frac{e^{ik_z z}J_0(\lambda\rho)}{\lambda^2+k_z^2-k^2}\lambda d\lambda dk_z \tag{4.308}$$

PROBLEMS

4.1. Given the time-harmonic representation of $f(t)$ in (4.23), show that

$$\frac{df}{dt} = \text{Re}(i\omega F e^{i\omega t})$$

4.2. Verify the four relations for the real-part operator, given in (4.27)–(4.30). *Hint*: To prove the two relations for derivatives and integrals, begin with the basic definition of a derivative and a Riemann integral.

4.3. One of the important theorems of electromagnetic theory is the *principle of duality* [12],[13]. Using duality, make the necessary changes in (4.80)–(4.83) to obtain the fields produced by the magnetic sheet source

$$\mathbf{M}(z) = \hat{x} M_{s0} \delta(z)$$

where M_{s0} is a constant magnetic surface current density in volts/m.

4.4. From (4.72), the wavenumber with loss is given by

$$k = k_d \sqrt{1 - iS}$$

Show that the requirement $\text{Im}(k) < 0$ implies that $\text{Re}(k) > 0$. *Hint*: Write ik in terms of its real and imaginary parts, viz.

$$ik = \alpha + i\beta = ik_d \sqrt{1 - iS}$$

Solve for α and β by squaring both sides and discarding the extraneous root. Note that $\text{Im}(k) < 0$ implies $\text{Re}(\alpha) > 0$. From the sign of α, it is then possible to infer the sign of β.

4.5. In (4.98), we obtained

$$\int_{-\infty}^{\infty} \frac{e^{i\beta z}}{\beta^2 - k^2} d\beta = \frac{\pi}{ik} e^{-ikz}, \qquad z > 0$$

By contour integration and the calculus of residues, obtain the result for $z < 0$. *Hint*: In Example 4.1, we closed the contour on a semi-circle through the upper half of the β-plane. For $z < 0$, close the contour through the lower half of the β-plane.

4.6. An interesting variation [14] on the line source problem examined in Section 4.3 is the line source located at (x', y') parallel to the z-axis and polarized in the ρ-direction (Fig. 4-10). Such a source can be represented by

$$\mathbf{J} = \hat{\rho} I_0 \frac{\delta(\rho - \rho')}{\rho} \delta(\phi - \phi')$$

Show that the magnetic field radiated by this source is given by

$$H_z = -\frac{I_0}{4\rho'} \sum_{-\infty}^{\infty} n e^{in(\phi-\phi')} \begin{cases} H_n^{(2)}(k\rho') J_n(k\rho), & \rho < \rho' \\ H_n^{(2)}(k\rho) J_n(k\rho'), & \rho > \rho' \end{cases}$$

Show that, despite the presence in the sum of the multiplicative factor n, the series converges as $n \to \pm\infty$.

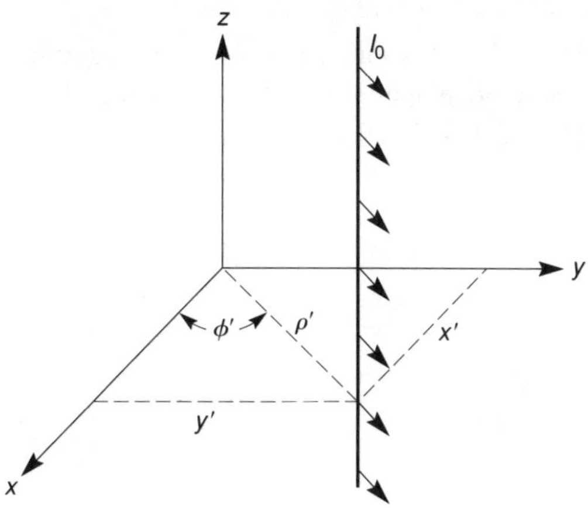

Fig. 4-10 Line source parallel to z-axis and ρ-polarized.

REFERENCES

[1] Friedman, B. (1956), *Principles and Techniques of Applied Mathematics*. New York: Wiley, 290–293.

[2] Collin, R.E. (1991), *Field Theory of Guided Waves*, 2nd edition. New York: IEEE Press, 2–3.

[3] Shen, L.C., and J.A. Kong (1987), *Applied Electromagnetism*, 2nd edition. Boston: PWS Engineering, chapter 1.

[4] Harrington, R.F. (1961, reissued 1987), *Time-Harmonic Electromagnetic Fields*. New York: McGraw-Hill, 16–18.

[5] Ibid., 106–116.

[6] Paul, C.R., and S.A. Nasar (1987), *Introduction to Electromagnetic Fields*. New York: McGraw-Hill, 298–305.

[7] Ibid.

[8] Gradshteyn, I.S., and I.M. Ryzhik (1980), *Table of Integrals, Series, and Products*. San Diego: Academic, 678, #6.532(4).

[9] Felsen, L.B., and N. Marcuvitz (1973), *Radiation and Scattering of Waves*. Englewood Cliffs, NJ: Prentice-Hall, 323–328.

[10] Ibid., 639–645.

[11] op.cit. Harrington, 227–228.

[12] op.cit. Harrington, 98–100.

[13] Balanis, C.A. (1989), *Advanced Engineering Electromagnetics*. New York: Wiley, 310–312.

[14] Butler, C.M. (1991), Vector potential Green's function for a radially-directed line dipole in open space, *Microwave and Optical Technology Letters* 4(11): 430–433.

5

Electromagnetic Boundary Value Problems

5.1 INTRODUCTION

In this chapter, we consider electromagnetic boundary value problems. We apply the concepts developed in the first four chapters, concentrating on the application of the mathematical ideas to a representative set of electromagnetic examples. Our principal objective is *structure*. Once the reader understands how the concepts in linear spaces, coupled with the theories of Green's functions and spectral expansions, can be applied to examples, it should become apparent how the methods are used to approach the study of electromagnetic propagation, scattering, and diffraction in an organized, logical manner.

We begin by extending the Green's function method to three dimensions. We next consider the case where the three-dimensional geometry is independent of one spatial coordinate so that the problem reduces to two dimensions. We then present a series of examples. We shall find that we may construct solutions in two and three dimensions by using combinations of the one-dimensional Green's functions and one-dimensional spectral representations discussed in Chapters 2 and 3, respectively.

One of the important solution characteristics that emerges in the examples is the fact that there exist alternative representations for the solutions. In particular, we exhibit alternative representations for the fields in a parallel plate waveguide and the fields scattered by a perfectly conducting cylinder. These solutions not only have important physical interpretations, but also are useful in different portions of the frequency spectrum.

5.2 SLP1 EXTENSION TO THREE DIMENSIONS

We begin by considering the negative Laplacian operator $L = -\nabla^2$ on a three-dimensional closed and bounded region V. This region is surrounded by a surface S whose parts may or may not be contiguous. For example (Fig. 5-1), the surface S might consist of an external surface S_e and two internal surfaces S_1 and S_2 with

$$S = S_e + S_1 + S_2$$

In this case, the region V consists of the volume internal to S_e but external to S_1 and S_2. By convention, the unit normal vector $\hat{n}$ points outward from V. Our interest is in the three-dimensional partial differential equation

$$L_\lambda u = f \tag{5.1}$$

where f is a real function and where

$$L_\lambda = L - \lambda = -\nabla^2 - \lambda, \qquad \lambda \in \mathbf{R} \tag{5.2}$$

The functional dependence of u and f is

$$u = u(\mathbf{r})$$

$$f = f(\mathbf{r})$$

where $\mathbf{r} \in V$. In Cartesian coordinates, for example, $u(\mathbf{r})$ stands for $u(x, y, z)$. Let $u(\mathbf{r})$ and $v(\mathbf{r})$ be members of a Hilbert space $\mathcal{H}$ with inner product

$$\langle u, v \rangle = \int_V u(\mathbf{r})v(\mathbf{r})dV \tag{5.3}$$

for all $u, v \in \mathcal{H}$. The three-dimensional problem involving (5.1) can be stated as follows: Given the partial differential equation in (5.1) and a suitable boundary condition involving $u(\mathbf{r})$ and its normal derivative on the surface S, determine $u(\mathbf{r})$ throughout V. We shall require that $u(\mathbf{r})$ have the following specification on S:

$$B(u) = \alpha_1 u\big|_S + \alpha_2 \nabla u\big|_S \cdot \hat{n} = \alpha \tag{5.4}$$

where the coefficients $\alpha, \alpha_1, \alpha_2$ are real and where $\hat{n}$ is the outgoing normal from the surface S. Our notation $u\big|_S$ indicates the function $u(\mathbf{r})$ evaluated at points on S. In addition, $\nabla u\big|_S \cdot \hat{n}$ indicates the normal derivative of $u(\mathbf{r})$ evaluated at points on S. The condition in (5.4) has two important special cases. If $\alpha_2 = 0$ and $\alpha_1 = 1$, we have

$$B(u) = u\big|_S = \alpha \tag{5.5}$$

Equation (5.1), coupled with boundary condition (5.5), is called the *Dirichlet problem*, and (5.5) is referred to as the *inhomogeneous Dirichlet boundary condition* [1]. (If $\alpha = 0$, the boundary condition is *homogeneous.*) If $\alpha_1 = 0$ and $\alpha_2 = 1$, we have

$$B(u) = \nabla u\big|_S \cdot \hat{n} = \alpha \tag{5.6}$$

Equation (5.1), coupled with boundary condition (5.6), is called the *Neumann problem*, and (5.6) is referred to as the *inhomogeneous Neumann boundary condition*. The general case in (5.4) is called the *mixed problem.* We point out that it is perfectly reasonable to have one type of boundary condition (Dirichlet, Neumann, or mixed) on a portion of the surface S and a different type of boundary condition (Dirichlet, Neumann, or mixed) on the remainder.

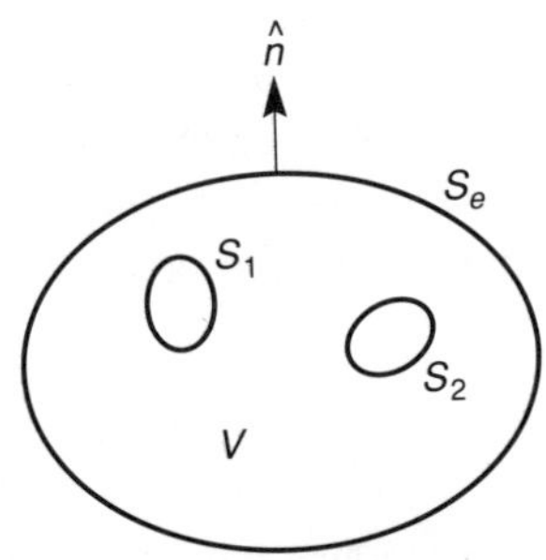

Fig. 5-1 Three-dimensional region V surrounded by surface S. In this case, $S = S_e + S_1 + S_2$.

We recognize the above collection of problems as a three-dimensional extension to SLP1, considered in Section 2.4. Most of the concepts developed in one dimension in Section 2.4 carry over to the present case, as we shall now demonstrate.

The operator $L = -\nabla^2$ has a formal adjoint. For $u, v \in \mathcal{H}$, we form

$$\langle Lu, v\rangle = \int_V (-\nabla^2 u) v \, dV \tag{5.7}$$

In the one-dimensional case, the adjoint was found by integrating by parts twice. The extension to three dimensions can be obtained by integrating by parts over all three coordinates comprising V. A more direct and compact method, however, employs *Green's theorem* [2]. In the case of the Laplacian operator, Green's theorem is given by

$$\int_V (-\nabla^2 u) v dV = \int_V u(-\nabla^2 v) dV + \int_S (-v\nabla u + u\nabla v) \cdot \hat{n} dS \tag{5.8}$$

We write this result in inner product notation as

$$\langle Lu, v \rangle = \langle u, L^* v \rangle + J(u, v) \Big|_S \tag{5.9}$$

where the conjunct $J(u, v) \Big|_S$ is given by

$$J(u, v) \Big|_S = \int_S (-v\nabla u + u\nabla v) \cdot \hat{n} dS \tag{5.10}$$

The operator L^* produced by Green's theorem in (5.9) is the formal adjoint to L. We observe in (5.8) that

$$L^* = L \tag{5.11}$$

and therefore, the operator $L = -\nabla^2$ is formally self-adjoint.

As was the case in Chapter 2, we shall assume initially that the boundary condition on u is homogeneous, $B(u) = 0$. We now choose the boundary condition on v to be that condition $B^*(v) = 0$ which, when coupled with the boundary condition on u, results in the vanishing of the conjunct, viz.

$$J(u, v) \Big|_S = 0 \tag{5.12}$$

In general, the boundary conditions associated with v are different from those associated with u. When they are the same, however, the operator L is self-adjoint.

EXAMPLE 5.1 Consider the homogeneous Dirichlet problem. Then, $B(u) = u\big|_S = 0$, and (5.12) gives

$$0 = \int_S v\nabla u \cdot \hat{n} dS$$

To satisfy this relationship, we choose $B^*(v) = v\big|_S = 0$. Since the boundary condition on v is identical to the boundary condition on u, the operator L for the Dirichlet problem is self-adjoint.

■

EXAMPLE 5.2 Consider the homogeneous Neumann problem. Then, $B(u) = \nabla u\big|_S \cdot \hat{n} = 0$, and (5.12) gives

$$0 = \int_S u \nabla v \cdot \hat{n} dS$$

To satisfy this relationship, we choose $B^*(v) = \nabla v\big|_S \cdot \hat{n} = 0$. Since the boundary condition on v is identical to the boundary condition on u, the operator L for the Neumann problem is self-adjoint.

■

To produce the solution to (5.1), we define two auxiliary problems: the Green's function problem and the adjoint Green's function problem. The Green's function problem is defined as follows:

$$L_\lambda g(\mathbf{r}, \mathbf{r}') = \delta(\mathbf{r} - \mathbf{r}') \tag{5.13}$$

$$B(g) = 0 \tag{5.14}$$

where L_λ is defined in (5.2). We note that, by definition, the boundary condition on g is identical to the homogeneous boundary condition on u. The adjoint Green's function problem is defined as follows:

$$L_\lambda h(\mathbf{r}, \mathbf{r}') = \delta(\mathbf{r} - \mathbf{r}') \tag{5.15}$$

$$B^*(h) = 0 \tag{5.16}$$

We note that, by definition, the boundary condition on h is identical to the boundary condition on v.

In the same manner as in Section 2.4, the solution to (5.1) is obtained by taking the inner product of $L_\lambda u$ with h, viz.

$$\langle L_\lambda u, h \rangle = \langle u, L_\lambda h \rangle + J(u, h)\Big|_S \tag{5.17}$$

where the integrations are with respect to the unprimed coordinates. Substitution of (5.1) and (5.15) into (5.17) gives

$$u(\mathbf{r}') = \langle f, h \rangle - J(u, h)\Big|_S \tag{5.18}$$

or, explicitly,

$$u(\mathbf{r}') = \int_V f(\mathbf{r}) h(\mathbf{r}, \mathbf{r}') dV + \int_S \left[h(\mathbf{r}, \mathbf{r}') \nabla u(\mathbf{r}) - u(\mathbf{r}) \nabla h(\mathbf{r}, \mathbf{r}') \right] \cdot \hat{n} dS \tag{5.19}$$

We note that (5.19) is the solution to (5.1), provided that we can determine the adjoint Green's function $h(\mathbf{r}, \mathbf{r}')$.

In a manner similar to the Green's function method developed in Section 2.4, we can show that it is never necessary to find the adjoint Green's function directly. Indeed, we form

$$\langle L_\lambda g(\mathbf{r}, \mathbf{r}'), h(\mathbf{r}, \mathbf{r}'')\rangle = \langle g(\mathbf{r}, \mathbf{r}'), L_\lambda h(\mathbf{r}, \mathbf{r}'')\rangle + J(g, h)\Big|_S \tag{5.20}$$

We are given the boundary conditions on g. We choose the boundary conditions on h so that

$$J(g, h)\Big|_S = 0 \tag{5.21}$$

Then, substitution of (5.13), (5.15), and (5.21) into (5.20) gives

$$h(\mathbf{r}', \mathbf{r}'') = g(\mathbf{r}'', \mathbf{r}')$$

or, with a change in variables,

$$h(\mathbf{r}, \mathbf{r}') = g(\mathbf{r}', \mathbf{r}) \tag{5.22}$$

Therefore, the adjoint Green's function is given simply by interchanging $\mathbf{r}$ and $\mathbf{r}'$ in the expression for the Green's function $g(\mathbf{r}, \mathbf{r}')$. In cases where L is self-adjoint, the boundary conditions on h are the same as those on g, and we must have

$$h(\mathbf{r}, \mathbf{r}') = g(\mathbf{r}, \mathbf{r}') = g(\mathbf{r}', \mathbf{r}) \tag{5.23}$$

Therefore, the Green's function is symmetric. For the self-adjoint case, we may substitute (5.23) into (5.19) to obtain

$$u(\mathbf{r}') = \int_V f(\mathbf{r}) g(\mathbf{r}, \mathbf{r}') dV + \int_S \left[g(\mathbf{r}, \mathbf{r}')\nabla u(\mathbf{r}) - u(\mathbf{r})\nabla g(\mathbf{r}, \mathbf{r}')\right] \cdot \hat{n} dS \tag{5.24}$$

For the case presently under consideration, where the boundary conditions on u are homogeneous, the term involving the conjunct in (5.24) vanishes, and we are left with

$$u(\mathbf{r}') = \int_V f(\mathbf{r}) g(\mathbf{r}, \mathbf{r}') dV \tag{5.25}$$

To extend the results to the inhomogeneous case, we simply apply the inhomogeneous boundary conditions to (5.24), with the result that some of the terms in the conjunct will survive. The final step in the solution involves the interchange of the primed and unprimed coordinates, in the same manner as in Section 2.4. We demonstrate these concepts in the following example.

EXAMPLE 5.3 Consider a rectangular box (Fig. 5-2) with dimensions a, b, c. It is required to find the solution to $-\nabla^2 u = f$ in the region V inside the box, where it is given that u satisfies the Dirichlet condition $B(u) = u\big|_S = 0$ on the boundary. The formulation of the problem is as follows:

$$-\nabla^2 u = f, \qquad \text{in } V \tag{5.26}$$

$$u(0, y, z) = u(a, y, z) = 0 \tag{5.27}$$

$$u(x, 0, z) = u(x, b, z) = 0 \tag{5.28}$$

$$u(x, y, 0) = u(x, y, c) = 0 \tag{5.29}$$

We know that the operator $L = -\nabla^2$ with Dirichlet boundary conditions is self-adjoint. We may therefore use (5.25) rather than (5.19), viz.

$$u(\mathbf{r}') = \int_V f(\mathbf{r})g(\mathbf{r}, \mathbf{r}')dV \tag{5.30}$$

where we require the solution to

$$\left(-\frac{\partial^2}{\partial x^2} - \frac{\partial^2}{\partial y^2} - \frac{\partial^2}{\partial z^2}\right) g(x, y, z, x', y', z') = \delta(x - x')\delta(y - y')\delta(z - z') \tag{5.31}$$

with

$$g\Big|_{x=0} = g\Big|_{x=a} = 0 \tag{5.32}$$

$$g\Big|_{y=0} = g\Big|_{y=b} = 0 \tag{5.33}$$

$$g\Big|_{z=0} = g\Big|_{z=c} = 0 \tag{5.34}$$

where we have chosen the boundary conditions on g to be identical to the boundary conditions on u. We begin the solution to (5.31) by invoking the spectral representation of $\delta(x - x')$. As we found in Problem 3.1, this representation produces the orthonormal eigenfunctions

$$u_m(x) = \sqrt{\frac{2}{a}} \sin \frac{m\pi x}{a} \tag{5.35}$$

and leads to the Fourier sine series, viz.

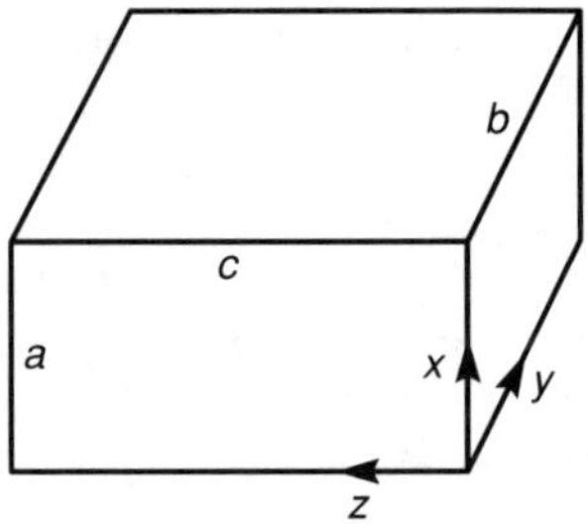

Fig. 5-2 Rectangular box problem.

$$g(x, y, z, x', y', z') = \sum_{m=1}^{\infty} \alpha_m(y, z, x', y', z') u_m(x) \tag{5.36}$$

where

$$\alpha_m(y, z, x', y', z') = \int_0^a g(x, y, z, x', y', z') u_m(x) dx \tag{5.37}$$

In a manner similar to (3.24) and (3.25), we note that (5.37) transforms the Green's function g into the coefficient α_m, viz.

$$g \Longrightarrow \alpha_m \tag{5.38}$$

Also, (5.36) provides the inverse transformation of α_m into g, viz.

$$g \Longleftarrow \alpha_m \tag{5.39}$$

Since the operator

$$L_x = -\frac{\partial^2}{\partial x^2}$$

is self-adjoint, (3.27) and (3.28) give

$$L_x g \Longrightarrow \left(\frac{m\pi}{a}\right)^2 \alpha_m \tag{5.40}$$

We also easily establish that

$$\int_0^a \delta(x - x') u_m(x) dx = u_m(x')$$

so that

$$\delta(x - x') \Longrightarrow u_m(x') \tag{5.41}$$

Using (5.38), (5.40), and (5.41) to transform (5.31), we obtain

$$-\left[\frac{\partial^2}{\partial y^2} + \frac{\partial^2}{\partial z^2} - \left(\frac{m\pi}{a}\right)^2\right] \alpha_m(y, z, x', y', z') = u_m(x')\delta(y - y')\delta(z - z') \tag{5.42}$$

Equation (5.42) is a partial differential equation whose solution yields α_m. To solve (5.42), we invoke the spectral representation of $\delta(y - y')$. The operator $(-\partial^2/\partial y^2)$ with boundary conditions

$$\alpha_m(0, z, x', y', z') = \alpha_m(b, z, x', y', z') = 0$$

results in orthonormal eigenfunctions that we may again obtain from the results in Problem 3.1, viz.

$$v_n(y) = \sqrt{\frac{2}{b}} \sin \frac{n\pi y}{b} \tag{5.43}$$

These eigenfunctions lead to a Fourier sine series representation of α_m, viz.

$$\alpha_m(y, z, x', y', z') = \sum_{n=1}^{\infty} \beta_{mn}(z, x', y', z') v_n(y) \tag{5.44}$$

where

$$\beta_{mn}(z, x', y', z') = \int_0^b \alpha_m(y, z, x', y', z') v_n(y) dy \tag{5.45}$$

Since the operator

$$L_y = -\frac{\partial^2}{\partial y^2}$$

is self-adjoint, we proceed in the same manner as in the above treatment of L_x. With respect to the transformation given in (5.45), we have

$$\alpha_m \Longrightarrow \beta_{mn}$$

$$L_y \alpha_m \Longrightarrow \left(\frac{n\pi}{b}\right)^2 \beta_{mn}$$

$$\delta(y - y') \Longrightarrow v_n(y')$$

We use these relations to transform (5.42), with the result

$$-\left(\frac{d^2}{dz^2} - \gamma_{mn}^2\right) \beta_{mn}(z, x', y', z') = u_m(x') v_n(y') \delta(z - z') \tag{5.46}$$

where

$$\gamma_{mn}^2 = \left(\frac{m\pi}{a}\right)^2 + \left(\frac{n\pi}{b}\right)^2 \tag{5.47}$$

We let

$$\beta_{mn}(z, x', y', z') = \hat{\beta}_{mn}(z, z') u_m(x') v_n(y') \tag{5.48}$$

and obtain the ordinary differential equation

$$\left(-\frac{d^2}{dz^2} + \gamma_{mn}^2\right)\hat{\beta}_{mn} = \delta(z - z') \tag{5.49}$$

We associate the following boundary conditions with (5.49):

$$\hat{\beta}_{mn}\bigg|_{z=0} = \hat{\beta}_{mn}\bigg|_{z=c} = 0 \tag{5.50}$$

It is easy to show that satisfaction of (5.50) satisfies (5.34). The Green's function problem posed by (5.49) with boundary conditions in (5.50) can be solved by the standard Green's function methods developed in Chapter 2. The details are left for the problems. The results are as follows:

$$\hat{\beta}_{mn}(z, z') = \frac{1}{\gamma_{mn}\sinh\gamma_{mn}c}\begin{cases}\sinh\gamma_{mn}(c - z')\sinh\gamma_{mn}z, & z < z' \\ \sinh\gamma_{mn}(c - z)\sinh\gamma_{mn}z' & z > z'\end{cases} \tag{5.51}$$

Substitution of (5.51) into (5.48), (5.48) into (5.44), and (5.44) into (5.36) yields the required Green's function, viz.

$$g = \sum_{m=1}^{\infty}\sum_{n=1}^{\infty}\frac{u_m(x)v_n(y)u_m(x')v_n(y')}{\gamma_{mn}\sinh\gamma_{mn}c}\cdot\begin{cases}\sinh\gamma_{mn}(c - z')\sinh\gamma_{mn}z, & z < z' \\ \sinh\gamma_{mn}(c - z)\sinh\gamma_{mn}z', & z > z'\end{cases} \tag{5.52}$$

We note in (5.52) that the Green's function is symmetric, $g(\mathbf{r}, \mathbf{r}') = g(\mathbf{r}', \mathbf{r})$, as predicted by the self-adjoint property of the negative Laplacian operator. Substitution of (5.52) into (5.30), followed by an interchange of the primed and unprimed coordinates, yields the final solution, viz.

$$\begin{aligned} u(x, y, z) = \sum_{m=1}^{\infty}\sum_{n=1}^{\infty}\frac{u_m(x)v_n(y)}{\gamma_{mn}\sinh\gamma_{mn}c}\int_0^b v_n(y')\int_0^a u_m(x')\Bigg[\sinh\gamma_{mn}(c - z)\int_0^z \\ \cdot f(x', y', z')\sinh\gamma_{mn}z'dz' + \sinh\gamma_{mn}z \\ \cdot\int_z^c f(x', y', z')\sinh\gamma_{mn}(c - z')dz'\Bigg]dx'dy' \end{aligned} \tag{5.53}$$

In the production of the Green's function, we chose to begin with a spectral expansion over the x-coordinate, followed by a spectral expansion over the y-coordinate. In a manner similar to many of the multiple-dimension cases considered in Chapter 4, alternate representations are possible. Other forms could be obtained by expanding spectrally over y and z or over x and z. Another possibility is to expand spectrally over all three coordinates.

■

5.3 SLP1 IN TWO DIMENSIONS

In many of the interesting problems in electromagnetics, the assumption is made that the fields are independent of one of the three spatial coordinates, with the result that the problem to be solved is two-dimensional. To solve two-dimensional problems, we modify the Green's function method developed for three dimensions in the previous section. The starting point is again the application of Green's theorem to the negative Laplacian operator, as in (5.8), viz.

$$\int_V (-\nabla^2 u)v\,dV = \int_V u(-\nabla^2 v)\,dV + \int_S (-v\nabla u + u\nabla v)\cdot \hat{n}\,dS \quad (5.54)$$

In the two-dimensional case, the Laplacian is two-dimensional. For example, if the problem is independent of z, we would write the negative Laplacian as $-\nabla_{xy}^2$. In addition, the volume V and the surface S become *degenerate*, in the sense that the integration over one of the coordinates involved in both the volume and the surface integrals is trivial. For example, if the problem is independent of z, the integrations over z in all three integrals in (5.54) will cancel. We illustrate these ideas in the following example.

EXAMPLE 5.4 Consider a rectangular cylinder (Fig. 5-3) with cross-sectional dimensions a and b. It is required to find the solution to $-\nabla_{xy}^2 u = f$ in the region V interior to the cylinder, where it is assumed that f is independent of z. Since the geometry is also independent of z, the solution u will be z-independent. Suppose it is given that homogeneous Dirichlet boundary conditions apply on the surfaces bounding the cylinder, except for the surface at $y = b$, where it is given that the inhomogeneous boundary condition

$$u|_{y=b} = \alpha$$

applies, where α is a real constant. We state the problem as follows:

$$-\nabla_{xy}^2 u = f \quad (5.55)$$

$$u(0, y) = u(a, y) = u(x, 0) = 0 \quad (5.56)$$

$$u(x, b) = \alpha \quad (5.57)$$

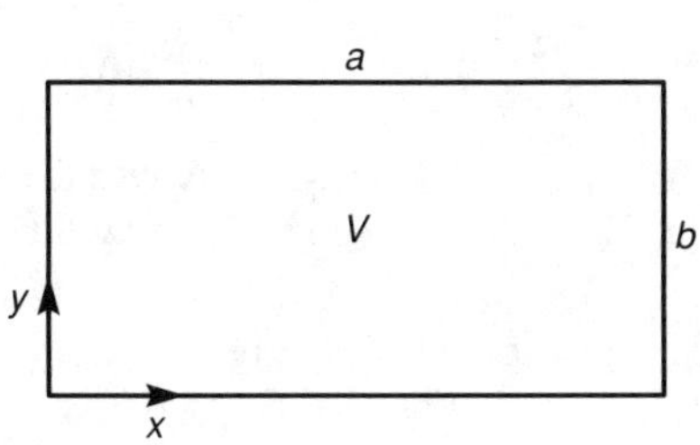

Fig. 5-3 Rectangular cylinder problem.

The associated Green's function problem is as follows:

$$-\nabla_{xy}^2 g = \delta(x - x')\delta(y - y') \tag{5.58}$$

$$g|_{x=0} = g|_{x=a} = g|_{y=0} = g|_{y=b} = 0 \tag{5.59}$$

where, consistent with the discussion associated with (5.13) and (5.14), we have chosen the boundary conditions associated with g to be the same as the homogeneous form of the boundary conditions associated with u. The homogeneous Dirichlet boundary condition case associated with (5.55) is a self-adjoint problem. We therefore formulate in terms of the Green's function $g(x, y, x', y')$. In this case, (5.54) becomes

$$\int_V (-\nabla^2 u) g dV = \int_V u(-\nabla^2 g) dV + \int_S (-g\nabla u + u\nabla g) \cdot \hat{n} dS \tag{5.60}$$

The surface integral S consists of integrals over the surfaces bounding the cylinder at $x = 0$, $x = a$, $y = 0$, and $y = b$, plus the integrals over the cylinder cross sections at $z \to -\infty$ and $z \to \infty$. Application of the boundary conditions specified in (5.56) and (5.59) reduces all surface integrals to zero, except over the surface at $y = b$ and over the surfaces at $z \to -\infty$ and $z \to \infty$. Consider the integral at $z \to \infty$. We have

$$\nabla u \cdot \hat{n} = \frac{\partial u}{\partial z}$$

$$\nabla g \cdot \hat{n} = \frac{\partial g}{\partial z}$$

By hypothesis, there are no variations with respect to z, and these partial derivatives vanish. The integral at $z \to -\infty$ vanishes in a similar manner. The only remaining surface integral contribution is over the surface at $y = b$, and (5.60) reduces to the following:

$$\int_{-\infty}^{\infty}\int_0^b\int_0^a (-\nabla^2 u) g dx dy dz = \int_{-\infty}^{\infty}\int_0^b\int_0^a u(-\nabla^2 g) dx dy dz + \alpha \int_{-\infty}^{\infty}\int_0^a (\nabla g \cdot \hat{y})\Big|_{y=b} dx dz \tag{5.61}$$

Since the integrands in all three integrals are independent of z, the z-integrations cancel and we have

$$\int_0^b\int_0^a (-\nabla_{xy}^2 u) g dx dy = \int_0^b\int_0^a u(-\nabla_{xy}^2 g) dx dy + \alpha \int_0^a \frac{\partial g}{\partial y}\Big|_{y=b} dx \tag{5.62}$$

where we have used $\partial/\partial z = 0$ to reduce the Laplacian to two dimensions. Substituting (5.55) and (5.58) and performing the delta function integration, we obtain

$$u(x', y') = \int_0^b\int_0^a f(x, y) g(x, y, x', y') dx dy - \alpha \int_0^a \frac{\partial g(x, b, x', y')}{\partial y} dx \tag{5.63}$$

To complete the solution, we must solve the Green's function problem given in (5.58) and (5.59). In the same manner as in Example 5.3, the spectral representation in the x-direction leads to a Fourier sine series, viz.

$$g(x, y, x', y') = \sum_{m=1}^{\infty} \alpha_m(y, x', y') u_m(x) \tag{5.64}$$

$$\alpha_m(y, x', y') = \int_0^a g(x, y, x', y') u_m(x) dx \tag{5.65}$$

where

$$u_m(x) = \sqrt{\frac{2}{a}} \sin \frac{m\pi x}{a} \tag{5.66}$$

With respect to the transformation given in (5.65), we have

$$g \Longrightarrow \alpha_m$$

Using the same procedure as in (5.37)–(5.41), we obtain

$$-\frac{\partial^2 g}{\partial x^2} \Longrightarrow \left(\frac{m\pi}{a}\right)^2 \alpha_m$$

$$\delta(x - x') \Longrightarrow u_m(x')$$

Using these relations to transform (5.58), we obtain

$$\left[-\frac{d^2}{dy^2} + \left(\frac{m\pi}{a}\right)^2\right] \alpha_m(y, x', y') = u_m(x')\delta(y - y') \tag{5.67}$$

We associate the following boundary conditions with (5.67):

$$\alpha_m(0, x', y') = \alpha_m(b, x', y') = 0 \tag{5.68}$$

Application of these Dirichlet boundary conditions satisfies the boundary condition requirements in (5.59) at $y = 0$ and $y = b$. The solution to (5.67) is available immediately from the result previously obtained in (5.51), viz.

$$\alpha_m = \frac{u_m(x')}{\frac{m\pi}{a} \sinh \frac{m\pi b}{a}} \begin{cases} \sinh \frac{m\pi(b-y')}{a} \sinh \frac{m\pi y}{a}, & y < y' \\ \sinh \frac{m\pi(b-y)}{a} \sinh \frac{m\pi y'}{a}, & y > y' \end{cases} \tag{5.69}$$

Substitution in (5.64) yields the Green's function

$$g(x, y, x', y') = \frac{2}{a} \sum_{m=1}^{\infty} \frac{\sin \frac{m\pi x}{a} \sin \frac{m\pi x'}{a}}{\frac{m\pi}{a} \sinh \frac{m\pi b}{a}} \begin{cases} \sinh \frac{m\pi(b-y')}{a} \sinh \frac{m\pi y}{a}, & y < y' \\ \sinh \frac{m\pi(b-y)}{a} \sinh \frac{m\pi y'}{a}, & y > y' \end{cases} \tag{5.70}$$

Equation (5.70) gives the Green's function required in the first integral in (5.63). In the second integral in (5.63), we require $\partial g/\partial y$ evaluated at $y = b$. Performing the required differentiation in (5.70) yields

$$\frac{\partial g(x, b, x', y')}{\partial y} = -\frac{2}{a}\sum_{m=1}^{\infty} \sin\frac{m\pi x}{a} \sin\frac{m\pi x'}{a} \frac{\sinh\frac{m\pi y'}{a}}{\sinh\frac{m\pi b}{a}} \tag{5.71}$$

Substitution of (5.70) and (5.71) into (5.63), followed by an interchange of the prime and unprimed coordinates, yields the following result:

$$u(x, y) = \alpha \int_0^a \frac{2}{a}\sum_{m=1}^{\infty} \sin\frac{m\pi x}{a} \sin\frac{m\pi x'}{a} \frac{\sinh\frac{m\pi y}{a}}{\sinh\frac{m\pi b}{a}} dx' + \frac{2}{a}\int_0^b\int_0^a dx'dy' f(x', y')$$

$$\cdot \sum_{m=1}^{\infty} \frac{\sin\frac{m\pi x}{a} \sin\frac{m\pi x'}{a}}{\frac{m\pi}{a}\sinh\frac{m\pi b}{a}} \begin{cases} \sinh\frac{m\pi(b-y')}{a} \sinh\frac{m\pi y}{a}, & y < y' \\ \sinh\frac{m\pi(b-y)}{a} \sinh\frac{m\pi y'}{a}, & y > y' \end{cases} \tag{5.72}$$

It is easy to show that (5.72) satisfies the boundary conditions at $x = 0$, $x = a$, and $y = 0$. The details are left for the reader. We now show that (5.72) satisfies the inhomogeneous boundary condition at $y = b$ required by (5.57). Indeed, at $y = b$, the second term vanishes and we have

$$u(x, b) = \alpha \int_0^a \frac{2}{a}\sum_{m=1}^{\infty} \sin\frac{m\pi x}{a} \sin\frac{m\pi x'}{a} dx' = \alpha \int_0^a \delta(x - x')dx' = \alpha \tag{5.73}$$

where we have used the spectral representation of the delta function

$$\delta(x - x') = \frac{2}{a}\sum_{m=1}^{\infty} \sin\frac{m\pi x}{a} \sin\frac{m\pi x'}{a}$$

It is also important to show that the solution in (5.72) satisfies the original differential equation in (5.55). The details are left for the Problems.

■

5.4 SLP2 AND SLP3 EXTENSION TO THREE DIMENSIONS

In this section, we consider complex f, complex λ, and admit the possibility of unbounded regions. We produce an SLP2 and SLP3 extension to three dimensions. We again confine our attention to the three-dimensional negative Laplacian operator and consider the partial differential equation

$$L_\lambda u = f \tag{5.74}$$

where

$$L_\lambda = L - \lambda, \qquad \lambda \in \mathbf{C} \tag{5.75}$$

and where L is the Laplacian operator

$$L = -\nabla^2 \tag{5.76}$$

Define the three-dimensional inner product

$$\langle f, g \rangle = \int_V f(\mathbf{r})\overline{g}(\mathbf{r})dV \tag{5.77}$$

The Green's function problem associated with (5.74) is given by

$$L_\lambda g = \delta(\mathbf{r} - \mathbf{r}') \tag{5.78}$$

Extending our analysis in Sections 2.5 and 2.6, we solve (5.74) by taking the inner product with the adjoint Green's function $h(\mathbf{r}, \mathbf{r}')$, as follows:

$$\langle L_\lambda u, h \rangle = \int_V \left[\left(-\nabla^2 - \lambda\right) u\right] \overline{h} dV = \int_V (-\nabla^2 u)\overline{h} dV + \int_V (-\lambda u)\overline{h} dV \tag{5.79}$$

To produce the adjoint operator L_λ^* and the conjunct $J(u, h)$, we again use *Green's theorem*. In the case of the Laplacian operator, we have

$$\int_V (-\nabla^2 u)\overline{h} dV = \int_V u(-\nabla^2 \overline{h})dV + \int_S \left(-\overline{h}\nabla u + u\nabla\overline{h}\right) \cdot \hat{n} dS \tag{5.80}$$

where S is the surface bounding V and $\hat{n}$ is the unit normal to S in the direction outward from V. But,

$$\int_V u(-\nabla^2 \overline{h})dV = \int_V u(\overline{-\nabla^2 h})dV = \langle u, -\nabla^2 h \rangle \tag{5.81}$$

and

$$\int_V (-\lambda u)\overline{h} dV = \int_V u(\overline{-\overline{\lambda} h})dV = \langle u, -\overline{\lambda} h \rangle \tag{5.82}$$

Substituting (5.81) in (5.80) and then (5.80) and (5.82) into (5.79), we obtain

$$\langle L_\lambda u, h \rangle = \langle u, L_\lambda^* h \rangle + J(u, h) \Big|_S \tag{5.83}$$

where

$$L_\lambda^* u = -\nabla^2 - \overline{\lambda} \tag{5.84}$$

$$J(u,h)\Big|_S = -\int_S \left(\overline{h}\nabla u - u\nabla\overline{h}\right)\cdot\hat{n}\,dS \tag{5.85}$$

We note that (5.83) is the three-dimensional extension of (2.134). We may solve for u in (5.83) by considering the adjoint Green's function problem given by

$$L_\lambda^* h = \delta(\mathbf{r}-\mathbf{r}') \tag{5.86}$$

Substitution of (5.74) and (5.86) into (5.83) yields

$$u(\mathbf{r}') = \langle f, h\rangle - J(u,h)\Big|_S \tag{5.87}$$

or, explicitly,

$$u(\mathbf{r}') = \int_V f(\mathbf{r})\overline{h}(\mathbf{r},\mathbf{r}')dV + \int_S \left[\overline{h}(\mathbf{r},\mathbf{r}')\nabla u(\mathbf{r}) - u(\mathbf{r})\nabla\overline{h}(\mathbf{r},\mathbf{r}')\right]\cdot\hat{n}\,dS \tag{5.88}$$

We note that (5.88) is the solution to (5.74), provided that we can determine the conjugate adjoint Green's function $\overline{h}(\mathbf{r},\mathbf{r}')$. Taking the complex conjugate of both sides of (5.86), we obtain a partial differential equation for $\overline{h}$, viz.

$$\overline{L_\lambda^* h(\mathbf{r},\mathbf{r}')} = L_\lambda \overline{h}(\mathbf{r},\mathbf{r}') = \delta(\mathbf{r}-\mathbf{r}') \tag{5.89}$$

As in Chapter 2, we can show that it is never necessary to find the conjugate adjoint Green's function directly. Indeed, we form

$$\langle L_\lambda g(\mathbf{r},\mathbf{r}'), h(\mathbf{r},\mathbf{r}'')\rangle = \langle g(\mathbf{r},\mathbf{r}'), L_\lambda^* h(\mathbf{r},\mathbf{r}'')\rangle + J(g,h)\Big|_S \tag{5.90}$$

We are given the boundary conditions on g. We choose the boundary conditions on $\overline{h}$ so that

$$J(g,h)\Big|_S = 0 \tag{5.91}$$

Then, substitution of (5.78), (5.86), and (5.91) into (5.90) gives

$$\overline{h}(\mathbf{r}',\mathbf{r}'') = g(\mathbf{r}'',\mathbf{r}')$$

or, with a change in variables,

$$\overline{h}(\mathbf{r},\mathbf{r}') = g(\mathbf{r}',\mathbf{r}) \tag{5.92}$$

Therefore, the conjugate adjoint Green's function is given simply by interchanging $\mathbf{r}$ and $\mathbf{r}'$ in the expression for the Green's function $g(\mathbf{r}, \mathbf{r}')$. In cases where the Green's function is symmetric, $g(\mathbf{r}, \mathbf{r}') = g(\mathbf{r}', \mathbf{r})$ and

$$\overline{h}(\mathbf{r}, \mathbf{r}') = g(\mathbf{r}, \mathbf{r}') = g(\mathbf{r}', \mathbf{r}) \qquad \text{(symmetric case)} \tag{5.93}$$

We shall find that the Green's function is symmetric in many of the interesting cases to follow. It is certainly symmetric in cases where the operator L is self-adjoint. In addition, we shall find symmetry, as we have previously found in Chapter 2, when examining many problems containing limit points and limit circles. For the symmetric Green's function case, we may substitute (5.93) into (5.88) to obtain

$$u(\mathbf{r}') = \int_V f(\mathbf{r}) g(\mathbf{r}, \mathbf{r}') dV + \int_S [g(\mathbf{r}, \mathbf{r}') \nabla u(\mathbf{r}) - u(\mathbf{r}) \nabla g(\mathbf{r}, \mathbf{r}')] \cdot \hat{n} dS \tag{5.94}$$

We shall summarize the steps for solving (5.74) by the above-described extension to the one-dimensional Green's function method. We distinguish two cases, dependent on whether or not the Green's function is symmetric.

Nonsymmetric Green's Function Case

1. Write the solution in the form given by (5.88).
2. Substitute the boundary conditions for u on the surface S into (5.88).
3. Substitute the conjugate adjoint boundary conditions for $\overline{h}$ on the surface S into (5.88).
4. Solve the Green's function problem given by (5.78) with boundary conditions on S the same as the homogeneous form of the boundary conditions on u.
5. Obtain the conjugate adjoint Green's function $\bar{h}$ through (5.92) and substitute into (5.88).
6. Interchange the variables $\mathbf{r}$ and $\mathbf{r}'$ in (5.88).

Symmetric Green's Function Case

1. Write the solution in the form given by (5.94).
2. Substitute the boundary conditions for u on the surface S into (5.94).
3. Substitute the boundary conditions for g on the surface S into (5.94).

4. Solve the Green's function problem given by (5.78) with boundary conditions on S the same as the homogeneous form of the boundary conditions on u and substitute into (5.94).
5. Interchange the variables $\mathbf{r}$ and $\mathbf{r}'$ in (5.94).

5.5 THE PARALLEL PLATE WAVEGUIDE

In this section, we consider the propagation of electromagnetic waves in a parallel plate waveguide. We consider a waveguide of uniform cross section with no scattering objects.

Suppose that the waveguide is formed from two parallel, perfectly conducting plates (Fig. 5-4), separated by a distance d and extending from $-\infty$ to ∞ in the y-direction and the z-direction. Assume that the medium between the parallel plates is free space. We shall choose the source of the electromagnetic field to be independent of y. Since the structure is also independent of y, we must have

$$\frac{\partial}{\partial y} = 0 \tag{5.95}$$

We begin with Maxwell's curl equations, given in (4.55) and (4.56). If ϵ_0 is the permittivity of free space, we have

$$\nabla \times \mathbf{H} = \mathbf{J} + i\omega\epsilon_0 \mathbf{E} \tag{5.96}$$

$$\nabla \times \mathbf{E} = -\mathbf{M} - i\omega\mu_0 \mathbf{H} \tag{5.97}$$

We expand these two equations in Cartesian coordinates and group them into two sets as follows:

Set 1: TM_z

$$\frac{\partial H_y}{\partial z} = -J_x - i\omega\epsilon_0 E_x \tag{5.98}$$

$$\frac{\partial E_x}{\partial z} - \frac{\partial E_z}{\partial x} = -M_y - i\omega\mu_0 H_y \tag{5.99}$$

$$\frac{\partial H_y}{\partial x} = J_z + i\omega\epsilon_0 E_z \tag{5.100}$$

Set 2: TE_z

$$\frac{\partial E_y}{\partial z} = M_x + i\omega\mu_0 H_x \tag{5.101}$$

$$\frac{\partial H_x}{\partial z} - \frac{\partial H_z}{\partial x} = J_y + i\omega\epsilon_0 E_y \tag{5.102}$$

$$\frac{\partial E_y}{\partial x} = -M_z - i\omega\mu_0 H_z \tag{5.103}$$

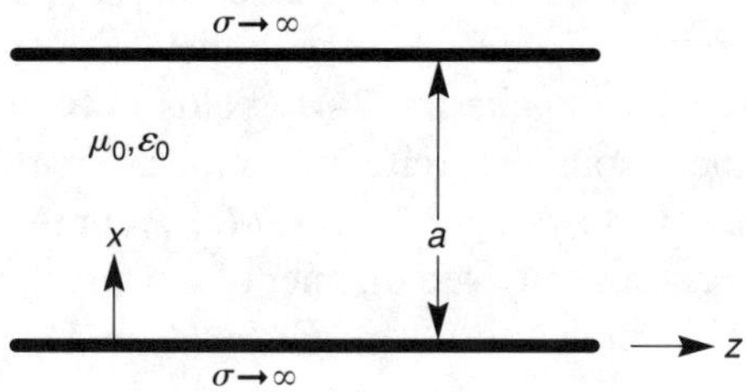

Fig. 5-4 Parallel plate waveguide.

We note that the transverse magnetic (TM_z) set is not coupled to the transverse electric (TE_z) set. It is therefore possible to excite one set independent of the other by appropriate selection of the **J** and **M** sources. We produce second-order partial differential equations governing each set by the following procedure. We differentiate (5.98) with respect to z, (5.100) with respect to x, add the result, and substitute (5.99) to obtain the following:

Set 1: TM_z

$$(\nabla_{xz}^2 + k^2)H_y = i\omega\epsilon_0 M_y + \frac{\partial J_z}{\partial x} - \frac{\partial Je_x}{\partial z} \tag{5.104}$$

$$E_x = -\frac{1}{i\omega\epsilon_0}\left(\frac{\partial H_y}{\partial z} + J_x\right) \tag{5.105}$$

$$E_z = \frac{1}{i\omega\epsilon_0}\left(\frac{\partial H_y}{\partial x} - J_z\right) \tag{5.106}$$

where

$$\nabla_{xz}^2 = \frac{\partial^2}{\partial x^2} + \frac{\partial^2}{\partial z^2} \tag{5.107}$$

and

$$k^2 = \omega^2\mu_0\epsilon_0 \tag{5.108}$$

A similar procedure applied to (5.101)–(5.103) yields

Set 2: TE_z

$$(\nabla_{xz}^2 + k^2)E_y = i\omega\mu_0 J_y - \frac{\partial M_z}{\partial x} + \frac{\partial M_x}{\partial z} \tag{5.109}$$

$$H_x = \frac{1}{i\omega\mu_0}\left(\frac{\partial E_y}{\partial z} - M_x\right) \tag{5.110}$$

$$H_z = -\frac{1}{i\omega\mu_0}\left(\frac{\partial E_y}{\partial x} + M_z\right) \tag{5.111}$$

We remark that the sources M_y, J_x, J_z excite the TM_z set, whereas the sources J_y, M_x, M_z excite the TE_z set. We shall consider the TM_z case.

To excite the TM_z fields only, we set $J_y = M_x = M_z = 0$ since these sources excite TE_z modes. We are left with three options to excite the TM_z modes, namely, M_y, J_x, or J_z. Since M_y enters into (5.104) in the least complicated manner of the three, we choose M_y and set $J_x = J_z = 0$. This choice results in $E_y = H_x = H_z = 0$ and

$$(\nabla_{xz}^2 + k^2)H_y = i\omega\epsilon_0 M_y \tag{5.112}$$

$$E_x = -\frac{1}{i\omega\epsilon_0}\frac{\partial H_y}{\partial z} \tag{5.113}$$

$$E_z = \frac{1}{i\omega\epsilon_0}\frac{\partial H_y}{\partial x} \tag{5.114}$$

At present, the only restriction on the source M_y is that it is independent of y. Associated with the Laplacian operator in (5.112) are the following boundary and limiting conditions on the field $H_y(x, z)$:

$$\lim_{z\to\pm\infty} H_y(x, z) = 0 \tag{5.115}$$

$$\frac{\partial H_y(0, z)}{\partial x} = \frac{\partial H_y(a, z)}{\partial x} = 0 \tag{5.116}$$

Equation (5.116) arises from the fact that E_z is tangential to the perfectly conducting waveguide surfaces at $x = 0$ and $x = a$, and therefore must vanish. Substituting this information in (5.114) yields (5.116). The Green's function problem associated with (5.112), (5.115), and (5.116) is as follows:

$$-(\nabla_{xz}^2 + k^2)g = \delta(x - x')\delta(z - z') \tag{5.117}$$

$$\lim_{z\to\pm\infty} g(x, z, x', z') = 0 \tag{5.118}$$

$$\frac{\partial g(0, z, x', z')}{\partial x} = \frac{\partial g(a, z, x', z')}{\partial x} = 0 \tag{5.119}$$

To solve for the TM_z fields, we shall solve (5.112) by the Green's function method given in Section 5.4. In the present case, however, the problem is independent of y. We introduce this simplification by beginning with (5.83), which gives, in this case,

$$\langle L_{k^2}H_y, h\rangle = \langle H_y, L_{k^2}^* h\rangle + J(H_y, h)\Big|_S \tag{5.120}$$

Explicitly,

$$\int_V \left[\left(-\nabla^2 - k^2\right) H_y\right] \overline{h} dV = \int_V H_y \left(-\nabla^2 - k^2\right) \overline{h} dV + \int_S \left(H_y \nabla \overline{h} - \overline{h} \nabla H_y\right) \cdot \hat{n} dS \tag{5.121}$$

We anticipate, and will verify, that the Green's function will be symmetric, and use (5.93) to write

$$\int_V \left[\left(-\nabla^2 - k^2\right) H_y\right] g dV = \int_V H_y \left(-\nabla^2 - k^2\right) g dV + \int_S \left(H_y \nabla g - g \nabla H_y\right) \cdot \hat{n} dS \tag{5.122}$$

The volume V is that region between the parallel plates. The surface S consists of the surfaces of the two parallel plates and the cross-sectional planar surfaces at $z \to \pm\infty$ and $y \to \pm\infty$. The surface integral contributions at $y \to \pm\infty$ vanish since

$$\nabla g \cdot \hat{n}\Big|_{y\to\pm\infty} = \pm \frac{\partial g}{\partial y}\Big|_{y\to\pm\infty} = 0$$

$$\nabla H_y \cdot \hat{n}\Big|_{y\to\pm\infty} = \pm \frac{\partial H_y}{\partial y}\Big|_{y\to\pm\infty} = 0$$

where we have used (5.95). The contributions on all remaining surfaces vanish with application of (5.115), (5.116), (5.118), and (5.119). The result is

$$\int_V \left[\left(-\nabla^2 - k^2\right) H_y\right] g dV = \int_V H_y \left(-\nabla^2 - k^2\right) g dV \tag{5.123}$$

Because of the absence of y-variations, the dy portion of the volume integrals cancel and the del-operator reduces to ∇_{xz}, with the result

$$\int_{-\infty}^{\infty} \int_0^a \left[\left(-\nabla_{xz}^2 - k^2\right) H_y\right] g dx dz = \int_{-\infty}^{\infty} \int_0^a H_y \left(-\nabla_{xz}^2 - k^2\right) g dx dz \tag{5.124}$$

Substituting (5.112) and (5.117) and performing the right-side integration, we obtain

$$H_y(x', z') = -i\omega\epsilon_0 \int_A M_y(x, z) g(x, z, x', z') dx dz$$

where A indicates the area occupied by the source. An interchange of primed and unprimed coordinates gives

$$H_y(x, z) = -i\omega\epsilon_0 \int_A M_y(x', z')g(x, z, x', z')dx'dz' \tag{5.125}$$

where we have again assumed that the Green's function is symmetric. Equation (5.125) gives the H_y-field everywhere inside the parallel plates, provided we can solve for the Green's function g, which we consider next.

To solve for the Green's function, defined by (5.117)–(5.119), we expand in terms of the spectral representation over the x-coordinate, viz.

$$g(x, z, x', z') = \sum_{n=0}^{\infty} \beta_n(z, x', z')u_n(x) \tag{5.126}$$

where the normalized eigenfunction $u_n(x)$ is given by

$$u_n(x) = \sqrt{\frac{\epsilon_n}{a}} \cos\frac{n\pi x}{a} \tag{5.127}$$

and ϵ_n is Neumann's number. The coefficient β_n is given by

$$\beta_n(z, x', z') = \int_0^a g(x, z, x', z')u_n(x)dx \tag{5.128}$$

With respect to the transformation in (5.128), we have

$$g \Longrightarrow \beta_n$$

Using the procedure in (5.37)–(5.41), we have

$$-\frac{\partial^2 g}{\partial x^2} \Longrightarrow \left(\frac{n\pi}{a}\right)^2 \beta_n$$

$$\delta(x - x') \Longrightarrow u_n(x')$$

Using these relations to transform (5.117), we obtain

$$\left(\frac{d^2}{dz^2} + k_z^2\right) \beta_n(z, x', z') = -u_n(x')\delta(z - z') \tag{5.129}$$

where

$$k_z^2 = k^2 - \left(\frac{n\pi}{a}\right)^2 \tag{5.130}$$

Let

$$\gamma_n = \frac{\beta_n}{u_n(x')} \tag{5.131}$$

With this definition, (5.129) becomes

$$\left(\frac{d^2}{dz^2} + k_z^2\right)\gamma_n = -\delta(z - z') \tag{5.132}$$

We associate the following limiting conditions with (5.132):

$$\lim_{z\to\pm\infty} \gamma_n = 0 \tag{5.133}$$

These conditions are consistent with those in (5.118). The solution to this one-dimensional Green's function differential equation has been obtained previously in Example 2.20. Applied to (5.132), we find that

$$\gamma_n = \frac{e^{-ik_z|z-z'|}}{2ik_z}, \qquad \operatorname{Im}(k_z) < 0 \tag{5.134}$$

Substituting (5.134) into (5.131) and the result into (5.126) gives

$$g(x, z, x', z') = \sum_{n=0}^{\infty}\left(\frac{\epsilon_n}{a}\right)\frac{e^{-ik_z|z-z'|}}{2ik_z}\cos\frac{n\pi x}{a}\cos\frac{n\pi x'}{a} \tag{5.135}$$

We note that the Green's function is symmetric, as anticipated. Substituting this result into (5.125) gives the magnetic field H_y, viz.

$$H_y(x, z) = -i\omega\epsilon_0 \sum_{n=0}^{\infty}\left[\int_A M_y(x', z')\left(\frac{\epsilon_n}{a}\right)\frac{e^{-ik_z|z-z'|}}{2ik_z}\cos\frac{n\pi x'}{a}dx'dz'\right]\cos\frac{n\pi x}{a} \tag{5.136}$$

The electric fields E_x and E_z associated with (5.136) can be calculated from (5.113) and (5.114), respectively.

We examine the modal structure of (5.136) by specializing the source M_y as follows:

$$M_y(x', z') = M_{sy}(x')\delta(z' - \ell) \tag{5.137}$$

where M_{sy} is a magnetic surface current in volts/m. Substituting (5.137) into (5.136) and performing the indicated z'-integration, we obtain

$$H_y(x, z) = -i\omega\epsilon_0 \sum_{n=0}^{\infty} B_n \frac{e^{-ik_z|z-\ell|}}{2ik_z}\sqrt{\frac{\epsilon_n}{a}}\cos\frac{n\pi x}{a} \tag{5.138}$$

where B_n is a *modal coefficient*, given by

$$B_n = \int_0^a M_{sy}(x')\sqrt{\frac{\epsilon_n}{a}} \cos\frac{n\pi x'}{a} dx' \tag{5.139}$$

The representation of the H_y-field in (5.138) shows the decomposition of the field into the familiar TEM $(n = 0)$ and TM $(n > 0)$ modes described in the undergraduate texts ([3],[4], for example). By making different choices of M_{sy}, we may adjust the coefficient B_n associated with each mode. For example, we may excite only the TEM mode by choosing

$$M_{sy} = \sqrt{\frac{1}{a}} M_0 \tag{5.140}$$

where M_0 is a constant. This choice gives

$$B_n = \begin{cases} 0, & n \neq 0 \\ M_0, & n = 0 \end{cases} \tag{5.141}$$

Substituting into (5.138), we obtain

$$H_y(x, z) = -i\omega\epsilon_0\sqrt{\frac{1}{a}} M_0 \frac{e^{-ik|z-\ell|}}{2ik_z} \tag{5.142}$$

which is a pure TEM wave traveling away from the source location $z = \ell$. For $z < \ell$, the wave travels right to left; for $z > \ell$, the wave travels left to right. If we are interested specifically in the region $z > \ell$, we may select the constant M_0 to produce a unit left-to-right TEM wave. Indeed, the choice

$$M_0 = \left(\frac{-i\omega\epsilon_0 e^{ik\ell}}{2i\sqrt{a}k_z}\right)^{-1} \tag{5.143}$$

produces

$$H_y(x, z) = e^{-ikz}, \qquad z > \ell \tag{5.144}$$

In (5.126), we chose to expand the Green's function in a spectral expansion over the x-coordinate. This expansion led to the Fourier cosine series, and produced a solution for the magnetic field H_y in terms of the waveguide modes. An *alternative representation* is possible. We shall begin by expanding the Green's function in a spectral expansion over z, rather than x. In (5.117), the spectral representation of the delta function associated with $-\partial^2/\partial z^2$ subject to the limiting condition in (5.118) leads to the Fourier transform pair

$$G(x, k_z, x', z') = \int_{-\infty}^{\infty} g(x, z, x', z') e^{-ik_z z} dz \tag{5.145}$$

$$g(x, z, x', z') = \frac{1}{2\pi} \int_{-\infty}^{\infty} G(x, k_z, x', z') e^{ik_z z} dk_z \tag{5.146}$$

We represent the transform pair by

$$g(x, z, x', z') \Longleftrightarrow G(x, k_z, x', z') \tag{5.147}$$

and easily find that

$$-\frac{\partial^2 g}{\partial z^2} \Longleftrightarrow k_z^2 G \tag{5.148}$$

$$\delta(z - z') \Longleftrightarrow e^{-ik_z z'} \tag{5.149}$$

Applying (5.145) and (5.147)–(5.149) to (5.117), we obtain

$$-\left(\frac{d^2}{dx^2} + k_x^2\right) G = e^{-ik_z z'} \delta(x - x') \tag{5.150}$$

where

$$k_x = \left(k^2 - k_z^2\right)^{1/2} \tag{5.151}$$

We let

$$\hat{G} = G e^{ik_z z'} \tag{5.152}$$

and obtain

$$-\left(\frac{d^2}{dx^2} + k_x^2\right) \hat{G} = \delta(x - x') \tag{5.153}$$

with boundary conditions inferred from (5.119), viz.

$$\left.\frac{d\hat{G}}{dx}\right|_{x=0} = \left.\frac{d\hat{G}}{dx}\right|_{x=a} = 0 \tag{5.154}$$

The solution to this differential equation and associated boundary conditions has been obtained previously in Example 2.13, viz.

$$\hat{G} = -\frac{1}{k_x \sin k_x a} \begin{cases} \cos k_x x \cos k_x (a - x'), & x < x' \\ \cos k_x x' \cos k_x (a - x) & x > x' \end{cases} \tag{5.155}$$

Using (5.152) and taking the inverse Fourier transform, we have

$$g(x,z,x',z') = -\frac{1}{2\pi}\int_{-\infty}^{\infty}\frac{e^{ik_z(z-z')}dk_z}{k_x \sin k_x a} \cdot \begin{cases} \cos k_x x \cos k_x(a-x'), & x < x' \\ \cos k_x x' \cos k_x(a-x) & x > x' \end{cases} \tag{5.156}$$

This representation of the Green's function is an alternative to the representation given in (5.135), which we repeat here for convenience, viz.

$$g(x,z,x',z') = \sum_{n=0}^{\infty}\left(\frac{\epsilon_n}{a}\right)\frac{e^{-ik_z|z-z'|}}{2ik_z}\cos\frac{n\pi x}{a}\cos\frac{n\pi x'}{a} \tag{5.157}$$

Although these two representations lead to the same Green's function $g(x,z,x',z')$, their forms are quite different. In (5.157), the cross-sectional waveguide modes are displayed explicitly. In (5.156), we find no explicit modal display. However, (5.156) is the starting point for constructing waveguide *ray representations* [5]. These ray representations are particularly useful in cases where the frequency is so high that a large number of modes can propagate in the waveguide. In addition, Felsen and Kamel have shown that the Green's function forms in (5.156) and (5.157) can be combined to produce what are called *hybrid ray-mode formulations*. The hybrid forms effectively exhibit the useful features in both the modal and ray formulations. The details can be found in [5].

In this section, we have studied formulations describing the propagation of the TEM and TM modes in a parallel plate waveguide. We leave the production of a modal series describing the TE modes for the problems. In the next section, we shall consider an obstacle in a parallel plate waveguide. We shall demonstrate the decomposition of the fields into incident, transmitted, and reflected waves.

5.6 IRIS IN PARALLEL PLATE WAVEGUIDE

In Section 5.5, we solved for the fields in a parallel plate waveguide. In this section, we add an iris to the waveguide interior. The waveguide has cross-sectional dimension a (Fig. 5-5), and contains an infinitesimally thin, perfectly conducting iris connected to the top plate at $x = a, z = 0$ and extending perpendicular to the plate and into the waveguide interior. The iris effectively divides the interior of the waveguide into two regions: Region 1, $z < 0$; Region 2, $z > 0$. The regions are connected electromagnetically

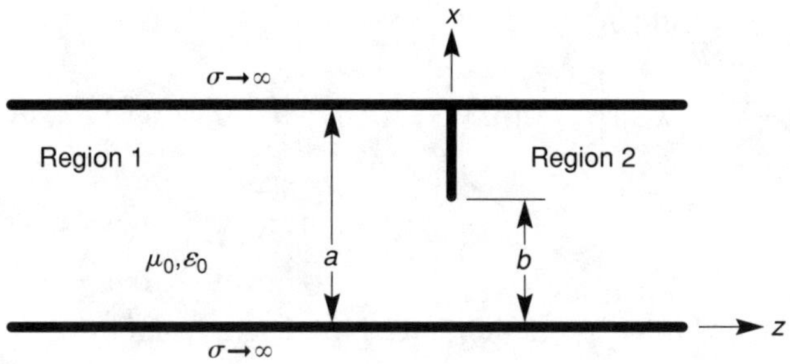

Fig. 5-5 Iris in parallel plate waveguide.

through an *aperture*, located at $z = 0, x \in (0, b)$. We note that the insertion of the iris does not change the y-independence of the waveguide geometry. Therefore, Maxwell's equations again separate into a TE_z set and a TM_z set, in the same manner as in Section 5.5. We shall cause excitation of the TM_z set by placing a constant magnetic sheet current source at $z = -d$ in Region 1, viz.

$$M_y = M_0 \delta(z + d) \tag{5.158}$$

Such a choice will produce a *TEM* wave incident from left to right in Region 1. As we shall discover in the ensuing analysis, this *TEM* wave, when encountering the iris, will cause *TEM* and *TM* waves to scatter from the iris into both Region 1 and Region 2. Using (5.112)–(5.114), we write the equations describing the TM_z fields in Region 1 and Region 2 as follows:

Region 1:

$$(\nabla_{xz}^2 + k^2) H_{y1} = i\omega\epsilon_0 M_y \tag{5.159}$$

$$E_{x1} = -\frac{1}{i\omega\epsilon_0} \frac{\partial H_{y1}}{\partial z} \tag{5.160}$$

$$E_{z1} = \frac{1}{i\omega\epsilon_0} \frac{\partial H_{y1}}{\partial x} \tag{5.161}$$

Region 2:

$$(\nabla_{xz}^2 + k^2) H_{y2} = 0 \tag{5.162}$$

$$E_{x2} = -\frac{1}{i\omega\epsilon_0} \frac{\partial H_{y2}}{\partial z} \tag{5.163}$$

$$E_{z2} = \frac{1}{i\omega\epsilon_0} \frac{\partial H_{y2}}{\partial x} \tag{5.164}$$

We first consider Region 1. Using the Green's function method and anticipating that the Green's function will be symmetric, we adapt (5.122) to the present case as follows:

$$\int_{V_1} \left[\left(-\nabla^2 - k^2\right) H_{y1}\right] g_1 dV = \int_{V_1} H_{y1} \left(-\nabla^2 - k^2\right) g_1 dV + \int_{S_1} (H_{y1} \nabla g_1 - g_1 \nabla H_{y1}) \cdot \hat{n} dS \tag{5.165}$$

where g_1 is the Green's function in Region 1, governed by

$$-(\nabla_{xz}^2 + k^2) g_1 = \delta(x - x')\delta(z - z') \tag{5.166}$$

with boundary and limiting conditions yet to be determined. The volume V_1 consists of Region 1. The surface S_1 consists of the following parts:

1. The surfaces of the two parallel plates in Region 1 ($z < 0$, $x = 0$ and $z < 0$, $x = a$).
2. The cross-sectional planar surfaces at $z \to -\infty$, $x \in (0, a)$ and $y \to \pm\infty$, $x \in (0, a)$.
3. The surface of the iris and aperture, $z = 0$, $x \in (0, a)$.

By the same reasoning as in the previous section, the surface integrals at $y \to \pm\infty$ vanish. In addition, we have the following boundary and limiting conditions governing the H_{y1}-fields:

$$\lim_{z \to -\infty} H_{y1} = 0 \tag{5.167}$$

$$\frac{\partial H_{y1}(0, z)}{\partial x} = \frac{\partial H_{y1}(a, z)}{\partial x} = 0 \tag{5.168}$$

$$\frac{\partial H_{y1}(x, 0)}{\partial z} = 0, \qquad x \in (b, a) \tag{5.169}$$

We choose the following boundary and limiting conditions for the Green's function g_1:

$$\lim_{z \to -\infty} g_1(x, z, x', z') = 0 \tag{5.170}$$

$$\frac{\partial g_1(0, z, x', z')}{\partial x} = \frac{\partial g_1(a, z, x', z')}{\partial x} = 0 \tag{5.171}$$

$$\frac{\partial g_1(x, 0, x', z')}{\partial z} = 0, \qquad x \in (0, a) \tag{5.172}$$

We note that the Green's function boundary and limiting conditions are the same as those associated with the H_{y1}-field, with one important exception. In (5.169), the boundary condition on H_{y1} is over the iris, whereas in

(5.172), the boundary condition is over the iris *and* the aperture. We therefore have a Green's function problem (Fig. 5-6) for a parallel plate waveguide extending from $z \to -\infty$ and terminating in a perfect conductor at $z = 0$. Before commenting on this choice, we substitute (5.167)–(5.172) into (5.165) and obtain

$$\int_{V_1} \left[\left(-\nabla^2 - k^2\right) H_{y1}\right] g_1 dV = \int_{V_1} H_{y1} \left(-\nabla^2 - k^2\right) g_1 dV - \int_{-\infty}^{\infty}\int_0^b \left(g_1 \frac{\partial H_{y1}}{\partial z}\right)\bigg|_{z=0} dxdy \tag{5.173}$$

Because of the invariance with y, the integrations with respect to y cancel in all terms. In addition, the del-operator reduces to ∇_{xz} and we have

$$\int_{-\infty}^{0}\int_0^a \left[\left(-\nabla_{xz}^2 - k^2\right) H_{y1}\right] g_1 dxdz = \int_{-\infty}^{0}\int_0^a H_{y1} \left(-\nabla_{xz}^2 - k^2\right) g_1 dxdz - \int_0^b \left(g_1 \frac{\partial H_{y1}}{\partial z}\right)\bigg|_{z=0} dx \tag{5.174}$$

Substituting (5.159) and (5.166), doing the delta-function integrations, and rearranging, we obtain

$$H_{y1}(x', z') = -i\omega\epsilon_0 \int_A M_y(x, z) g_1(x, z, x', z') dxdz + \int_0^b g_1(x, 0, x', z') \frac{\partial H_{y1}(x, 0)}{\partial z} dx$$

Anticipating the symmetry of the Green's function, we interchange the primed and unprimed coordinates and then substitute (5.158) and (5.160) to obtain

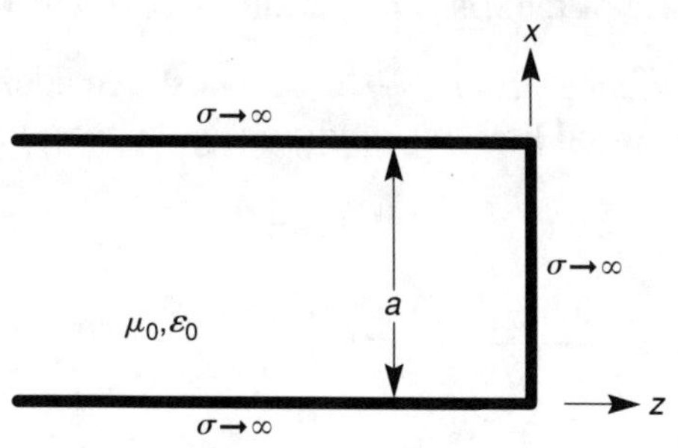

Fig. 5-6 Green's function problem for Region 1.

$$H_{y1}(x,z) = -i\omega\epsilon_0 M_0 \int_0^a g_1(x,z,x',-d)dx' - i\omega\epsilon_0 \cdot \int_0^b g_1(x,z,x',0)E_{x1}(x',0)dx' \tag{5.175}$$

Our choice of the boundary condition in (5.172) deserves some comment. If we had chosen this condition to be the same as that governing the H-field derivative in (5.169), we would have produced a Green's function problem with an unspecified boundary condition over $x \in (0,b)$, $z = 0$. Our choice completes the specification of the Green's function in Region 1, and has the added advantage of simplifying the problem solution by eliminating a portion of the surface integral.

We now consider Region 2. In a manner similar to our treatment in Region 1, we obtain

$$\int_{V_2} \left[\left(-\nabla^2 - k^2\right) H_{y2}\right] g_2 dV = \int_{V_2} H_{y2}\left(-\nabla^2 - k^2\right) g_2 dV + \int_{S_2} (H_{y2}\nabla g_2 - g_2\nabla H_{y2}) \cdot \hat{n} dS \tag{5.176}$$

where g_2 is the Green's function in Region 2, governed by

$$-(\nabla_{xz}^2 + k^2)g_2 = \delta(x-x')\delta(z-z') \tag{5.177}$$

The volume V_2 consists of Region 2. The surface S_2 consists of the following parts:

1. The surfaces of the two parallel plates in Region 2 ($z > 0$, $x = 0$ and $z > 0$, $x = a$).
2. The cross-sectional planar surfaces at $z \to \infty$, $x \in (0,a)$ and $y \to \pm\infty$, $x \in (0,a)$.
3. The surface of the iris and aperture, $z = 0$, $x \in (0,a)$.

Again, the surface integrals at $y \to \pm\infty$ vanish. In addition, we have the following boundary and limiting conditions governing the H_{y2}-fields:

$$\lim_{z\to\infty} H_{y2} = 0 \tag{5.178}$$

$$\frac{\partial H_{y2}(0,z)}{\partial x} = \frac{\partial H_{y2}(a,z)}{\partial x} = 0 \tag{5.179}$$

$$\frac{\partial H_{y2}(x,0)}{\partial z} = 0, \qquad x \in (b,a) \tag{5.180}$$

We choose the following boundary and limiting conditions for the Green's function g_2:

$$\lim_{z\to\infty} g_2(x, z, x', z') = 0 \tag{5.181}$$

$$\frac{\partial g_2(0, z, x', z')}{\partial x} = \frac{\partial g_2(a, z, x', z')}{\partial x} = 0 \tag{5.182}$$

$$\frac{\partial g_2(x, 0, x', z')}{\partial z} = 0, \qquad x \in (0, a) \tag{5.183}$$

We therefore have a Green's function problem (Fig. 5-7) for a parallel plate waveguide extending to $z \to \infty$ and terminating in a perfect conductor at $z = 0$. We substitute (5.178)–(5.183) into (5.176) and obtain

$$\int_{V_2} \left[\left(-\nabla^2 - k^2\right) H_{y2}\right] g_2 dV = \int_{V_2} H_{y2}\left(-\nabla^2 - k^2\right) g_2 dV + \int_{-\infty}^{\infty}\int_0^b \left(g_2 \frac{\partial H_{y2}}{\partial z}\right)\bigg|_{z=0} dxdy \tag{5.184}$$

Again, the integrations with respect to y cancel in all terms and the del-operator reduces to ∇_{xz}. We have

$$\int_0^{\infty}\int_0^a \left[\left(-\nabla_{xz}^2 - k^2\right) H_{y2}\right] g_2 dxdz = \int_0^{\infty}\int_0^a H_{y2}\left(-\nabla_{xz}^2 - k^2\right) g_2 dxdz + \int_0^b \left(g_2 \frac{\partial H_{y2}}{\partial z}\right)\bigg|_{z=0} dx \tag{5.185}$$

Substituting (5.162) and (5.177), doing the delta-function integrations, and rearranging, we obtain

$$H_{y2}(x', z') = -\int_0^b g_2(x, 0, x', z')\frac{\partial H_{y2}(x, 0)}{\partial z} dx$$

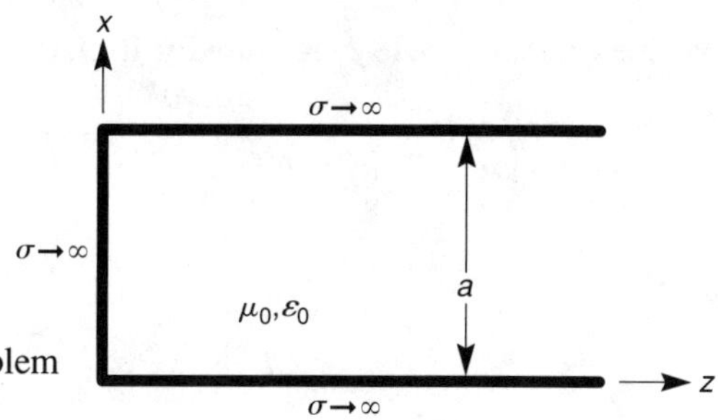

Fig. 5-7 Green's function problem for Region 2.

Anticipating the symmetry of the Green's function, we interchange the primed and unprimed coordinates and substitute (5.163) to obtain

$$H_{y2}(x, z) = i\omega\epsilon_0 \int_0^b g_2(x, z, x', 0) E_{x2}(x', 0) dx' \tag{5.186}$$

We next consider the determination of the Green's functions. In Region 1, we wish to solve (5.166) with boundary conditions given by (5.170)–(5.172). As we found in Example 3.2, the spectral representation of $(-\partial^2/\partial x^2)$ with Neumann boundary conditions given in (5.171) results in the Fourier cosine series. We therefore expand the Green's function

$$g_1(x, z, x', z') = \sum_{n=0}^{\infty} \alpha_n(z, x', z') u_n(x) \tag{5.187}$$

where $u_n(x)$ is the orthonormal eigenfunction

$$u_n(x) = \sqrt{\frac{\epsilon_n}{a}} \cos \frac{n\pi x}{a} \tag{5.188}$$

and where

$$\alpha_n(z, x', z') = \int_0^a g_1(x, z, x', z') u_n(x) dx \tag{5.189}$$

Equation (5.189) defines the transformation

$$g_1 \Longrightarrow \alpha_n \tag{5.190}$$

Using the procedure in (5.37)–(5.41), we establish that

$$-\frac{\partial^2 g_1}{\partial x^2} \Longrightarrow \left(\frac{n\pi}{a}\right)^2 \alpha_n \tag{5.191}$$

$$\delta(x - x') \Longrightarrow u_n(x') \tag{5.192}$$

We use (5.190)–(5.192) to transform (5.166), with the result

$$-\left(\frac{d^2}{dz^2} + k_z^2\right) \alpha_n(z, x', z') = u_n(x') \delta(z - z') \tag{5.193}$$

where

$$k_z^2 = k^2 - \left(\frac{n\pi}{a}\right)^2 \tag{5.194}$$

We define

$$\beta_n(z, x', z') = \frac{\alpha_n(z, x', z')}{u_n(x')} \tag{5.195}$$

and obtain

$$-\left(\frac{d^2}{dz^2} + k_z^2\right) \beta_n(z, x', z') = \delta(z - z') \tag{5.196}$$

We assign the boundary and limiting conditions

$$\lim_{z \to -\infty} \beta_n(z, x', z') = 0 \tag{5.197}$$

$$\frac{d\beta_n(0, x', z')}{dz} = 0, \qquad x \in (0, a) \tag{5.198}$$

Satisfaction of these two conditions results in the satisfaction of the conditions in (5.170) and (5.172). The differential equation in (5.196) with conditions in (5.197) and (5.198) is solved by the standard Green's function methods developed in Chapter 2. The details are left for the problems. The result is

$$\beta_n(z, x', z') = \frac{1}{ik_z} \begin{cases} e^{ik_z z} \cos k_z z', & z < z' \\ e^{ik_z z'} \cos k_z z, & z > z' \end{cases} \tag{5.199}$$

where

$$\mathrm{Im}(k_z) < 0$$

Substituting (5.199) into (5.195), solving for α_n, and substituting the result into (5.187) gives

$$g_1(x, z, x', z') = \sum_{n=0}^{\infty} \frac{\epsilon_n}{ik_z a} \cos \frac{n\pi x}{a} \cos \frac{n\pi x'}{a} \begin{cases} e^{ik_z z} \cos k_z z', & z < z' \\ e^{ik_z z'} \cos k_z z, & z > z' \end{cases} \tag{5.200}$$

To obtain the Green's function g_2 for Region 2, we note that the geometry in Fig. 5-7 can be obtained from Fig. 5-6 by reflection through the $z = 0$ plane. We therefore can obtain g_2 from g_1 by replacing z by $-z$ and z' by $-z'$, with the result

$$g_2(x, z, x', z') = \sum_{n=0}^{\infty} \frac{\epsilon_n}{ik_z a} \cos \frac{n\pi x}{a} \cos \frac{n\pi x'}{a} \begin{cases} e^{-ik_z z} \cos k_z z', & z > z' \\ e^{-ik_z z'} \cos k_z z, & z < z' \end{cases} \tag{5.201}$$

We note that the anticipated symmetry of the Green's functions occurs for both g_1 and g_2. In the solution for the fields, we shall need the Green's functions with $z' = 0$. We obtain

$$g_1(x, z, x', 0) = \sum_{n=0}^{\infty} \frac{\epsilon_n}{ik_z a} e^{ik_z z} \cos\frac{n\pi x}{a} \cos\frac{n\pi x'}{a} \tag{5.202}$$

$$g_2(x, z, x', 0) = \sum_{n=0}^{\infty} \frac{\epsilon_n}{ik_z a} e^{-ik_z z} \cos\frac{n\pi x}{a} \cos\frac{n\pi x'}{a} \tag{5.203}$$

We also need the Green's function in Region 1 evaluated with $z' = -d$. Confining our interest to regions to the right of the source, we obtain

$$g_1(x, z, x', -d) = \sum_{n=0}^{\infty} \frac{\epsilon_n}{ik_z a} \cos\frac{n\pi x}{a} \cos\frac{n\pi x'}{a} e^{-ik_z d} \cos k_z z, \qquad -d < z < 0 \tag{5.204}$$

We note that in (5.202)–(5.204), care must be taken to select the proper cases for the Green's functions g_1 and g_2. In (5.202), $z < z'$; in (5.203), $z > z'$; and, in (5.204), $z > z' = -d$.

The magnetic current source in Region 1, given by (5.158), has been chosen to produce a *TEM* wave incident from left to right. We may show this by substituting (5.204) into the first term in (5.175) to give

$$H_{y1}(x, z) = -\frac{M_0}{\eta} e^{-ikd} \cos kz - i\omega\epsilon_0 \int_0^b g_1(x, z, x', 0) E_{x1}(x', 0) dx' \tag{5.205}$$

As a normalization, we choose

$$M_0 = -2e^{ikd} \tag{5.206}$$

and obtain

$$H_{y1}(x, z) = \frac{2\cos kz}{\eta} - i\omega\epsilon_0 \int_0^b g_1(x, z, x', 0) E_{x1}(x', 0) dx' \tag{5.207}$$

This normalization is used to produce a unit-strength incident electric field. Indeed, consider the limiting case where the size of the aperture shrinks to zero. We have

$$\lim_{b\to 0} H_{y1}(x, z) = \frac{1}{\eta}\left(e^{ikz} + e^{-ikz}\right)$$

Substituting into (5.160) gives

$$\lim_{b\to 0} E_{x1}(x, z) = e^{ikz} - e^{-ikz}$$

Therefore, in the limit, we produce a unit-strength electric field, incident from left to right and reflecting from a perfect conductor at $z = 0$.

As a final step in the problem formulation, we note in (5.186) and (5.207) that

$$E_{x1}(x', 0) = E_{x2}(x', 0), \qquad z' \in (0, b) \tag{5.208}$$

which is a statement that the tangential electric field is continuous in the aperture. We symbolize this aperture field by $E_A(x')$ and write (5.186) and (5.207) as

$$H_{y1}(x, z) = \frac{2\cos kz}{\eta} - i\omega\epsilon_0 \int_0^b g_1(x, z, x', 0)E_A(x')dx' \tag{5.209}$$

$$H_{y2}(x, z) = i\omega\epsilon_0 \int_0^b g_2(x, z, x', 0)E_A(x')dx' \tag{5.210}$$

Substitution of (5.202) and (5.203) gives

$$H_{y1}(x, z) = \frac{2\cos kz}{\eta} - \frac{k}{\eta}\sum_{n=0}^{\infty}\frac{\epsilon_n}{k_z a}e^{ik_z z}\cos\frac{n\pi x}{a}\int_0^b E_A(x')\cos\frac{n\pi x'}{a}dx' \tag{5.211}$$

$$H_{y2}(x, z) = \frac{k}{\eta}\sum_{n=0}^{\infty}\frac{\epsilon_n}{k_z a}e^{-ik_z z}\cos\frac{n\pi x}{a}\int_0^b E_A(x')\cos\frac{n\pi x'}{a}dx' \tag{5.212}$$

These two expressions give the magnetic fields everywhere in the two regions. We note, however, that the electric field E_A in the aperture is as yet unknown. We have, however, one additional boundary condition that we have not utilized, namely, the continuity of the tangential magnetic field in the aperture. We invoke this continuity by equating (5.211) and (5.212) in the aperture and obtain

$$1 = \sum_{n=0}^{\infty}\frac{\epsilon_n}{a}\frac{k}{k_z}\cos\frac{n\pi x}{a}\int_0^b E_A(x')\cos\frac{n\pi x'}{a}dx' \tag{5.213}$$

or

$$1 = \int_0^b E_A(x')\mathcal{G}(x, x')dx' \tag{5.214}$$

where

$$\mathcal{G}(x, x') = \sum_{n=0}^{\infty}\frac{\epsilon_n}{a}\frac{k}{k_z}\cos\frac{n\pi x}{a}\cos\frac{n\pi x'}{a} \tag{5.215}$$

Expression (5.214) is an integral equation whose solution yields the aperture field E_A. Once E_A is known, the result can be substituted into

(5.211) and (5.212) to yield the magnetic fields. The corresponding electric fields can be obtained by substitution of these results into (5.160), (5.161), (5.163), and (5.164).

Unfortunately, the integral equation in (5.214) cannot be solved analytically. A popular method for finding an approximate solution for the aperture field E_A is the Method of Moments (MOM), introduced in Section 1.8. Although the approximate solution to (5.214) is beyond the central theme of this book, a few comments are in order.

The kernel $\mathcal{G}(x, x')$ for the integral equation in (5.214) is logarithmically singular. Therefore, care must be taken in dealing with the limit as $x' \to x$. For a discussion of the issues involved, the reader is referred to [6]–[8]. The series contained in $\mathcal{G}(x, x')$ is slowly converging. For methods to speed the convergence, the reader is referred to [6]–[10]. Finally, the aperture field possesses an edge singularity [6],[11] in its behavior as $x \to b$. This singularity must be considered in evaluations involving MOM. For a discussion, the reader is referred to [6],[8],[12].

5.7 APERTURE DIFFRACTION

We next consider the classic problem of diffraction by an aperture in a perfectly conducting screen. A perfectly conducting screen (Fig. 5-8) divides unbounded empty space into two regions: Region 1, $y < 0$; Region 2, $y > 0$. An aperture interrupts the screen at $y = 0$, $x \in (-a/2, a/2)$. Electromagnetic fields are excited by a z-directed magnetic current source M_z in Region 1. It is assumed that the source and the geometry are z-independent, so that

$$\frac{\partial}{\partial z} = 0 \tag{5.216}$$

Expanding Maxwell's curl equations in Cartesian coordinates and invoking (5.216), we have

$$\frac{\partial H_z}{\partial y} = i\omega\epsilon_0 E_x \tag{5.217}$$

$$\frac{\partial H_z}{\partial x} = -i\omega\epsilon_0 E_y \tag{5.218}$$

$$\frac{\partial H_y}{\partial x} - \frac{\partial H_x}{\partial y} = i\omega\epsilon_0 E_z \tag{5.219}$$

$$\frac{\partial E_z}{\partial y} = -i\omega\mu_0 H_x \tag{5.220}$$

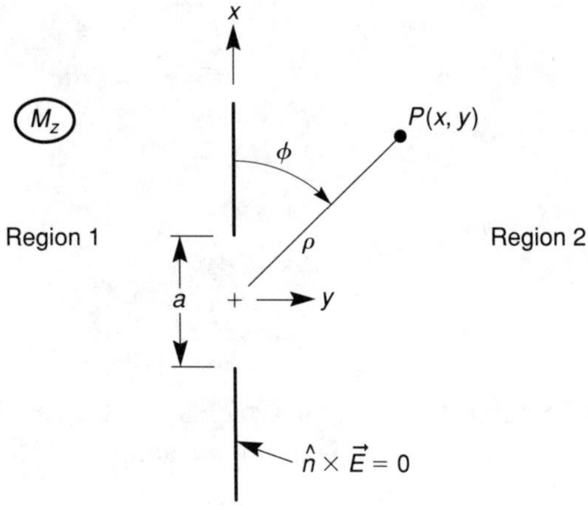

Fig. 5-8 Aperture in a perfectly conducting screen.

$$\frac{\partial E_z}{\partial x} = i\omega\mu_0 H_y \tag{5.221}$$

$$\frac{\partial E_x}{\partial y} - \frac{\partial E_y}{\partial x} = M_z + i\omega\mu_0 H_z \tag{5.222}$$

The set comprised of (5.217), (5.218), and (5.222) is excited by M_z and is decoupled from the unexcited set comprised of (5.219)–(5.221). We therefore have $H_x = H_y = E_z = 0$. Differentiating (5.217) with respect to y, (5.218) with respect to x, adding, and then substituting (5.222) gives the following set:

$$\left(\nabla_{xy} + k^2\right) H_z = i\omega\epsilon_0 M_z \tag{5.223}$$

$$E_x = \frac{1}{i\omega\epsilon_0}\frac{\partial H_z}{\partial y} \tag{5.224}$$

$$E_y = -\frac{1}{i\omega\epsilon_0}\frac{\partial H_z}{\partial x} \tag{5.225}$$

where k is given by (5.108). We next specialize the above equations to Regions 1 and 2 as follows:

Region 1:

$$\left(\nabla_{xy} + k^2\right) H_{z1} = i\omega\epsilon_0 M_z \tag{5.226}$$

$$E_{x1} = \frac{1}{i\omega\epsilon_0}\frac{\partial H_{z1}}{\partial y} \tag{5.227}$$

$$E_{y1} = -\frac{1}{i\omega\epsilon_0}\frac{\partial H_{z1}}{\partial x} \tag{5.228}$$

Region 2:

$$\left(\nabla_{xy} + k^2\right) H_{z2} = 0 \tag{5.229}$$

$$E_{x2} = \frac{1}{i\omega\epsilon_0}\frac{\partial H_{z2}}{\partial y} \tag{5.230}$$

$$E_{y2} = -\frac{1}{i\omega\epsilon_0}\frac{\partial H_{z2}}{\partial x} \tag{5.231}$$

We first consider Region 1. Using the Green's function method and anticipating symmetry of the Green's function, we adapt (5.165) to the present case as follows:

$$\begin{aligned}\int_{V_1} \left[\left(-\nabla^2 - k^2\right) H_{z1}\right] g_1 dV = \int_{V_1} H_{z1} \left(-\nabla^2 - k^2\right) g_1 dV \\ + \int_{S_1} \left(H_{z1}\nabla g_1 - g_1 \nabla H_{z1}\right) \cdot \hat{n} dS\end{aligned} \tag{5.232}$$

where g_1 is the Green's function in Region 1, governed by

$$-\left(\nabla_{xy}^2 + k^2\right) g_1 = \delta(x - x')\delta(y - y') \tag{5.233}$$

with boundary and limiting conditions to be determined. The volume V_1 is Region 1. The surface S_1 consists of the following parts:

1. The surface of the screen and aperture at $y = 0$.
2. The planar surface at $y \to -\infty$.
3. The planar surfaces at $x \to \pm\infty$, $y < 0$.
4. The planar surfaces at $z \to \pm\infty$, $y < 0$.

By the same reasoning as in the treatment of the parallel plate waveguide, the surface integrals at $z \to \pm\infty$ vanish. In addition, we have the following boundary and limiting conditions governing the H_{z1}-fields:

$$\lim_{y\to-\infty} H_{z1} = 0 \tag{5.234}$$

$$\lim_{x\to\pm\infty} H_{z1} = 0, \qquad y \in (-\infty, 0) \tag{5.235}$$

$$\frac{\partial H_{z1}(x,0)}{\partial y} = 0, \qquad x \notin (-a/2, a/2) \tag{5.236}$$

Inspection of (5.227) indicates that the condition in (5.236) is equivalent to the vanishing of the tangential electric field on the surface of the screen. We choose the following boundary and limiting conditions for the Green's function g_1:

$$\lim_{y\to-\infty} g_1(x, y, x', y') = 0 \tag{5.237}$$

$$\lim_{x\to\pm\infty} g_1(x, y, x', y') = 0, \qquad y \in (-\infty, 0) \tag{5.238}$$

$$\frac{\partial g_1(x, 0, x', y')}{\partial y} = 0 \tag{5.239}$$

We note that the Green's function boundary and limiting conditions are the same as those associated with the H_{z1}-field, with one exception. At $y = 0$, the boundary condition on H_{z1} is over the screen, whereas the boundary condition on g_1 is over the screen *and* the aperture. Therefore, the Green's function problem is for the half-space $y < 0$ with a perfectly conducting surface at $y = 0$. We substitute (5.234)–(5.239) into (5.232) and obtain

$$\int_{V_1} \left[\left(-\nabla^2 - k^2\right) H_{z1}\right] g_1 dV = \int_{V_1} H_{z1} \left(-\nabla^2 - k^2\right) g_1 dV - \int_{-\infty}^{\infty} \int_{-a/2}^{a/2} g_1 \frac{\partial H_{z1}}{\partial y} dx dz \tag{5.240}$$

Because of the invariance with z, the integrations with respect to z cancel in all terms. In addition, the del-operator reduces to ∇_{xy} and we have

$$\int_{-\infty}^{0} \int_{-\infty}^{\infty} \left[\left(-\nabla_{xy}^2 - k^2\right) dH_{z1}\right] g_1 dx dy = \int_{-\infty}^{0} \int_{-\infty}^{\infty} H_{z1} \left(-\nabla_{xy}^2 - k^2\right) g_1 dx dy - \int_{-a/2}^{a/2} g_1 \frac{\partial H_{z1}}{\partial y} dx \tag{5.241}$$

Substituting (5.226) and (5.233), doing the delta-function integrations, and rearranging, we obtain

$$H_{z1}(x', y') = -i\omega\epsilon_0 \int_A M_z(x, y) g_1(x, y, x', y') dx dy + \int_{-a/2}^{a/2} g_1(x, 0, x', y') \frac{\partial H_{z1}(x, 0)}{\partial y} dx$$

Anticipating the symmetry of the Green's function, we interchange the primed and unprimed coordinates and substitute (5.227) to obtain

$$H_{z1}(x, y) = -i\omega\epsilon_0 \int_A M_z(x', y')g_1(x, y, x', y')dx'dy'$$
$$+ i\omega\epsilon_0 \int_{-a/2}^{a/2} g_1(x, y, x', 0)E_{x1}(x', 0)dx' \tag{5.242}$$

We now consider Region 2. In a manner similar to Region 1, we obtain

$$\int_{V_2} \left[\left(-\nabla^2 - k^2\right) H_{z2}\right] g_2 dV = \int_{V_2} H_{z2} \left(-\nabla^2 - k^2\right) g_2 dV$$
$$+ \int_{S_2} (H_{z2}\nabla g_2 - g_2 \nabla H_{z2}) \cdot \hat{n} dS \tag{5.243}$$

where g_2 is the Green's function in Region 2, governed by

$$-\left(\nabla_{xy}^2 + k^2\right) g_2 = \delta(x - x')\delta(y - y') \tag{5.244}$$

with boundary and limiting conditions to be determined. The volume V_2 is Region 2. The surface S_2 consists of the following parts:

1. The surface of the screen and aperture at $y = 0$.
2. The planar surface at $y \to \infty$.
3. The planar surfaces at $x \to \pm\infty$, $y > 0$.
4. The planar surfaces at $z \to \pm\infty$, $y > 0$.

Again, the surface integrals at $z \to \pm\infty$ vanish. In addition, we have the following boundary and limiting conditions governing the H_{z2}-fields:

$$\lim_{y\to\infty} H_{z2} = 0 \tag{5.245}$$

$$\lim_{x\to\pm\infty} H_{z2} = 0, \qquad y \in (0, \infty) \tag{5.246}$$

$$\frac{\partial H_{z2}(x, 0)}{\partial y} = 0, \qquad x \notin (-a/2, a/2) \tag{5.247}$$

We choose the following boundary and limiting conditions for the Green's function g_2:

$$\lim_{y\to\infty} g_2(x, y, x', y') = 0 \tag{5.248}$$

$$\lim_{x\to\pm\infty} g_2(x, y, x', y') = 0, \qquad y \in (0, \infty) \tag{5.249}$$

$$\frac{\partial g_2(x, 0, x', y')}{\partial y} = 0 \tag{5.250}$$

We therefore have a Green's function problem for a half-space $y > 0$ with a perfectly conducting surface at $y = 0$. We substitute (5.245)–(5.250) into (5.243) and obtain

$$\int_{V_2} \left[\left(-\nabla^2 - k^2 \right) H_{z2} \right] g_2 dV = \int_{V_2} H_{z2} \left(-\nabla^2 - k^2 \right) g_2 dV + \int_{-\infty}^{\infty} \int_{-a/2}^{a/2} \left(g_2 \frac{\partial H_{z2}}{\partial y} \right) \Bigg|_{z=0} dx dz \tag{5.251}$$

Again, the integrations with respect to z cancel in all terms and the del-operator reduces to ∇_{xy}. We have

$$\int_0^{\infty} \int_{-\infty}^{\infty} \left[\left(-\nabla_{xz}^2 - k^2 \right) H_{z2} \right] g_2 dx dy = \int_0^{\infty} \int_{-\infty}^{\infty} H_{z2} \left(-\nabla_{xz}^2 - k^2 \right) g_2 dx dy + \int_{-a/2}^{a/2} \left(g_2 \frac{\partial H_{z2}}{\partial y} \right) \Bigg|_{z=0} dx \tag{5.252}$$

Substituting (5.229) and (5.244), doing the delta-function integrations, and rearranging, we obtain

$$H_{z2}(x', y') = -\int_{-a/2}^{a/2} g_2(x, 0, x', y') \frac{\partial H_{z2}(x, 0)}{\partial y} dx$$

Anticipating the symmetry of the Green's function, we interchange the primed and unprimed coordinates and substitute (5.230) to obtain

$$H_{z2}(x, y) = -i\omega\epsilon_0 \int_{-a/2}^{a/2} g_2(x, y, x', 0) E_{x2}(x', 0) dx' \tag{5.253}$$

At this point in the development, we have two remaining tasks, namely, the specification of boundary conditions at points in the aperture and the selection of a specific magnetic current source. We first require that the tangential electric field in the aperture be continuous. We symbolize the aperture electric field by $E_A(x)$ and write

$$E_{x1}(x, 0) = E_{x2}(x, 0) = E_A(x), \qquad x \in (-a/2, a/2) \tag{5.254}$$

We next specialize the source M_z to be a line source located at $x' = \xi$, $y' = \eta$, viz.

$$M_z(x', y') = M_0 \delta(x' - \xi) \delta(y' - \eta) \tag{5.255}$$

We substitute (5.254) into both (5.242) and (5.253). Also, we substitute (5.255) into (5.242) and perform the indicated delta-function integrations to give

$$H_{z1}(x,y) = -i\omega\epsilon_0 M_0 g_1(x,y,\xi,\eta) + i\omega\epsilon_0 \int_{-a/2}^{a/2} g_1(x,y,x',0) E_A(x') dx' \tag{5.256}$$

$$H_{z2}(x,y) = -i\omega\epsilon_0 \int_{-a/2}^{a/2} g_2(x,y,x',0) E_A(x') dx' \tag{5.257}$$

Expressions (5.256) and (5.257) give the magnetic fields everywhere, provided that we know the Green's functions g_1 and g_2 and provided we can find the aperture electric field E_A. We shall derive the Green's functions subsequently. The aperture field E_A is obtained by requiring the tangential magnetic field in the aperture to be continuous, viz.

$$H_{z1}(x,0) = H_{z2}(x,0), \qquad x \in (-a/2, a/2) \tag{5.258}$$

Substituting (5.256) and (5.257) into (5.258) yields the integral equation

$$M_0 g_1(x,0,\xi,\eta) = \int_{-a/2}^{a/2} \left[g_1(x,0,x',0) + g_2(x,0,x',0) \right] E_A(x') dx' \tag{5.259}$$

Once the Green's functions have been determined, an approximate solution to the integral equation (using Method of Moments, for example) yields an approximation to the aperture field E_A. The aperture field can then be substituted into (5.256) and (5.257) to give the magnetic fields everywhere. Once the magnetic fields are known, the electric fields can be obtained by differentiation in (5.227), (5.228), (5.230), and (5.231). We next consider the Green's functions g_1 and g_2.

We may determine g_2 from (5.244) and (5.248)–(5.250), which we reproduce here for convenience, viz.

$$\left(-\frac{\partial^2}{\partial x^2} - \frac{\partial^2}{\partial y^2} - k^2 \right) g_2 = \delta(x-x')\delta(y-y') \tag{5.260}$$

$$\lim_{x \to \pm\infty} g_2(x,y,x',y') = 0, \qquad y \in (0,\infty) \tag{5.261}$$

$$\frac{\partial g_2(x,0,x',y')}{\partial y} = 0 \tag{5.262}$$

$$\lim_{y \to \infty} g_2(x,y,x',y') = 0 \tag{5.263}$$

The spectral representation of $(-\partial^2/\partial x^2)$ with limiting conditions in (5.261) leads to the Fourier transform, as given in Example 3.4. In this case, we have

$$g_2(x, y, x', y') \Longleftrightarrow G_2(k_x, y, x', y') \tag{5.264}$$

$$-\frac{\partial^2 g_2}{\partial x^2} \Longleftrightarrow k_x^2 G_2 \tag{5.265}$$

$$\delta(x - x') \Longleftrightarrow e^{-ik_x x'} \tag{5.266}$$

Applying these relationships to (5.260) gives

$$\left(-\frac{d^2}{dy^2} - k_y^2\right) \hat{G}_2 = \delta(y - y') \tag{5.267}$$

where

$$k_y = \sqrt{k^2 - k_x^2} \tag{5.268}$$

$$G_2 = e^{-ik_x x'} \hat{G}_2 \tag{5.269}$$

The boundary and limiting conditions associated with (5.267) are as follows:

$$\frac{d\hat{G}_2(k_x, 0, x', y')}{dy} = 0 \tag{5.270}$$

$$\lim_{y \to \infty} \hat{G}_2(x, y, x', y') = 0 \tag{5.271}$$

Invoking these conditions is consistent with the conditions on g_2. The solution to (5.267) with the above associated conditions can be inferred from the Green's function problem discussed previously in (5.196)–(5.199). If in (5.199) we let $z \to -y$, $z' \to -y'$, and $k_z \to k_y$, we obtain

$$\hat{G}_2 = \frac{1}{ik_y} \begin{cases} e^{-ik_y y'} \cos k_y y, & y < y' \\ e^{-ik_y y} \cos k_y y', & y > y' \end{cases}$$

Expanding the cosine terms into exponentials gives the following useful alternate form:

$$\hat{G}_2 = \frac{1}{2ik_y} \left[e^{-ik_y|y-y'|} + e^{-ik_y(y+y')} \right] \tag{5.272}$$

Substituting this result into (5.269) and then taking the inverse Fourier transform, we have

$$g_2 = \frac{1}{4\pi i}\int_{-\infty}^{\infty} \frac{e^{-ik_y|y-y'|} + e^{-ik_y(y+y')}}{k_y} e^{ik_x(x-x')} dk_x \tag{5.273}$$

From (4.156), we have the following identity:

$$H_0^{(2)}\left(k\sqrt{x^2+y^2}\right) = \frac{1}{\pi}\int_{-\infty}^{\infty} \frac{e^{-ik_y|y|}e^{ik_x x}}{k_y} dk_x \tag{5.274}$$

Therefore,

$$\begin{aligned} g_2 = \frac{1}{4i}\Big\{ & H_0^{(2)}\left[k\sqrt{(x-x')^2+(y-y')^2}\right] \\ & + H_0^{(2)}\left[k\sqrt{(x-x')^2+(y+y')^2}\right]\Big\} \end{aligned} \tag{5.275}$$

Referring to the development in Section 4.6, we recognize (5.275) as a description of the radiation from a line source located at $x = x'$, $y = y'$, plus a line source at the image location $x = x'$, $y = -y'$, with respect to the ground plane.

Consider g_1. We may obtain the solution for g_1 directly from the above solution for g_2 by replacing y by $-y$ and y' by $-y'$. Such replacement does not change the solution, and we therefore have

$$g_2 = g_1 \tag{5.276}$$

The above determination of the Green's functions completes the formulation of the problem.

In many applications, the source M_z is located at a distance far enough from the aperture so that a plane wave approximation can be invoked. To accomplish this, we consider the Green's function in the first term in (5.256), viz.

$$\begin{aligned} g_1(x, y, \xi, \eta) = \frac{1}{4i}\Big\{ & H_0^{(2)}\left[k\sqrt{(x-\xi)^2+(y-\eta)^2}\right] \\ & + H_0^{(2)}\left[k\sqrt{(x-\xi)^2+(y+\eta)^2}\right]\Big\} \end{aligned} \tag{5.277}$$

Rewriting this expression in cylindrical coordinates, we have

$$\begin{aligned} g_1(\rho, \phi, \rho', \phi') = \frac{1}{4i}\Big\{ & H_0^{(2)}\left[k\sqrt{\rho^2+\rho'^2-2\rho\rho'\cos(\phi-\phi')}\right] \\ & + H_0^{(2)}\left[k\sqrt{\rho^2+\rho'^2-2\rho\rho'\cos(\phi+\phi')}\right]\Big\} \end{aligned} \tag{5.278}$$

where (ρ', ϕ') marks the position of the line source (Fig. 5-9). We may locate the line source at a distance remote from the aperture by letting ρ' become very large, in which case

$$\left[\rho^2 + \rho'^2 - 2\rho\rho' \cos(\phi \mp \phi')\right]^{1/2} \longrightarrow \rho' - \rho \cos(\phi \mp \phi') \qquad (5.279)$$

Furthermore, using the large argument approximation for the Hankel function, given in Example 2.21, we have

$$H_0^{(2)} \left\{k\left[\rho' - \rho\cos(\phi \mp \phi')\right]\right\} \sim \sqrt{\frac{2i}{\pi k \rho'}} e^{-ik\rho'} e^{ik\rho\cos(\phi \mp \phi')} \qquad (5.280)$$

Using (5.280) in (5.278), we obtain

$$\begin{aligned} g_1(\rho, \phi, \rho', \phi') &\sim \frac{1}{4i}\sqrt{\frac{2i}{\pi k \rho'}} e^{-ik\rho'} \left[e^{ik\rho\cos(\phi - \phi')} + e^{ik\rho\cos(\phi+\phi')}\right] \\ &= \frac{1}{4i}\sqrt{\frac{2i}{\pi k \rho'}} e^{-ik\rho'} 2 e^{ikx\cos\phi'} \cos(ky \sin\phi') \end{aligned} \qquad (5.281)$$

Let

$$M_0 = \left[\frac{-\omega\epsilon_0}{4}\sqrt{\frac{2i}{\pi k \rho'}} e^{-ik\rho'}\right]^{-1} \qquad (5.282)$$

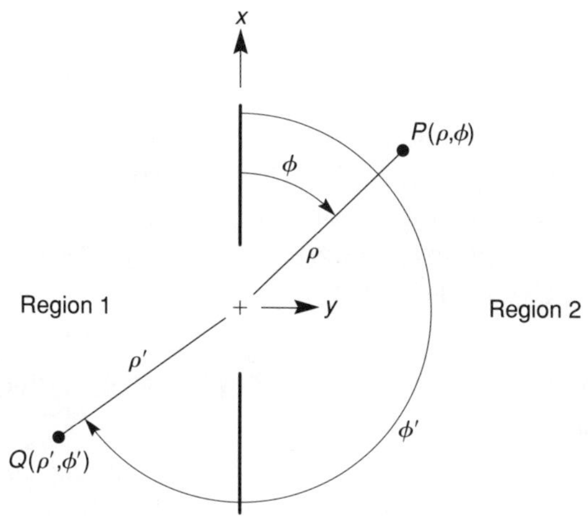

Fig. 5-9 Aperture in a perfectly conducting screen, cylindrical coordinates.

We substitute (5.275), (5.276), (5.281), and (5.282) into (5.256) and (5.257) and obtain

$$H_{z1}(x, y) = 2e^{ikx\cos\phi'}\cos(ky\sin\phi') + \frac{k}{2\eta}\int_{-a/2}^{a/2} H_0^{(2)}\left[k\sqrt{(x-x')^2+y^2}\right] E_A(x')dx' \tag{5.283}$$

$$H_{z2}(x, y) = -\frac{k}{2\eta}\int_{-a/2}^{a/2} H_0^{(2)}\left[k\sqrt{(x-x')^2+y^2}\right] E_A(x')dx' \tag{5.284}$$

where we have used

$$\omega\epsilon_0 = \frac{k}{\eta}$$

We note that in the limit as the aperture length approaches zero, we have

$$\begin{aligned}\lim_{a\to 0} H_{z1}(x, y) &= 2e^{ikx\cos\phi'}\cos(ky\sin\phi') \\ &= e^{ikx\cos\phi'}\left(e^{iky\sin\phi'} + e^{-iky\sin\phi'}\right)\end{aligned} \tag{5.285}$$

which represents a unit magnitude plane wave approaching the screen at angle ϕ' and reflecting according to Snell's law of reflection. Continuity of the tangential H-field in the aperture gives the integral equation

$$-\frac{2\eta}{k}e^{ikx\cos\phi'} = \int_{-a/2}^{a/2} H_0^{(2)}(k|x-x'|)E_A(x')dx' \tag{5.286}$$

which completes the problem formulation for the case of plane wave incidence.

Again, the integral equation in (5.286) cannot be inverted analytically. The aperture field E_A therefore must be determined approximately using numerical methods. We note, however, that for the aperture size a small enough, the Hankel function $H_0^{(2)}(k|x-x'|)$ can be approximated by $\ln|x-x'|$, in which case an analytical solution is possible in terms of Chebyshev polynomials. The resulting integral equation is considered in Problem 1.23. The details are given in [13],[14].

5.8 SCATTERING BY A PERFECTLY CONDUCTING CYLINDER

Consider an electric current J_z that excites a surface current on a perfectly conducting cylinder of uniform cross section (Fig. 5-10). We assume that the cylinder geometry and the source are independent of z, so that

$$\frac{\partial}{\partial z} = 0 \tag{5.287}$$

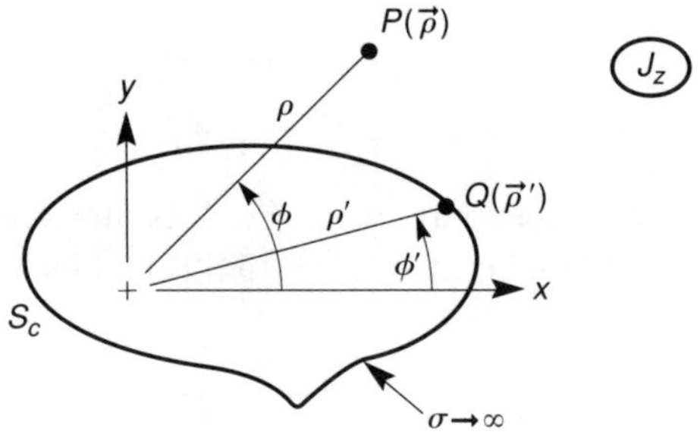

Fig. 5-10 Electric current J_z exciting a perfectly conducting cylinder.

We require the fields at a point $P(\rho)$. We begin with Maxwell's curl equations, viz.

$$\nabla \times \mathbf{H} = \hat{z}J_z + i\omega\epsilon_0\mathbf{E} \tag{5.288}$$

$$\nabla \times \mathbf{E} = -i\omega\mu_0\mathbf{H} \tag{5.289}$$

The curl operator in (5.288) and (5.289) is composed of components transverse to the z-direction, plus a z-component, viz.

$$\nabla = \nabla_t + \hat{z}\frac{\partial}{\partial z}$$

In this case, because of (5.287), we have

$$\nabla = \nabla_t \tag{5.290}$$

We divide the electric and magnetic fields into transverse and z-components, viz.

$$\mathbf{H} = \mathbf{H}_t + \hat{z}H_z \tag{5.291}$$

$$\mathbf{E} = \mathbf{E}_t + \hat{z}E_z \tag{5.292}$$

Substituting (5.290)–(5.292) into (5.288) and (5.289), we obtain

$$\nabla_t \times \mathbf{H}_t + \nabla_t \times \hat{z}H_z = \hat{z}J_z + i\omega\epsilon_0\mathbf{E} \tag{5.293}$$

$$\nabla_t \times \mathbf{E}_t + \nabla_t \times \hat{z}E_z = -i\omega\mu_0\mathbf{H} \tag{5.294}$$

Equating transverse components and z-components on either side of (5.293) and (5.294), we produce the following two sets:

Set 1: TM_z

$$\nabla_t \times \mathbf{H}_t = \hat{z}\,(J_z + i\omega\epsilon_0 E_z) \tag{5.295}$$

$$\nabla_t \times \hat{z}E_z = -i\omega\mu_0\mathbf{H}_t \tag{5.296}$$

Set 2: TE_z

$$\nabla_t \times \mathbf{E}_t = -\hat{z} i\omega\mu_0 H_z \tag{5.297}$$

$$\nabla_t \times \hat{z} H_z = i\omega\epsilon_0 \mathbf{E}_t \tag{5.298}$$

Set 1 is excited by the source J_z, while Set 2 is unexcited. We therefore have $E_t = H_z = 0$. We take the curl of (5.296) and substitute (5.295) to obtain

$$-i\omega\mu_0 \hat{z}\left(J_z + i\omega\epsilon_0 E_z\right) = \nabla_t \times \nabla_t \times \hat{z} E_z$$

$$= \left[\nabla_t\left(\nabla_t \cdot \hat{z} E_z\right) - \nabla_t^2(\hat{z} E_z)\right] \tag{5.299}$$

where we have used a well-known vector identity to expand the double-curl. But,

$$\nabla_t \cdot \hat{z} E_z = 0 \tag{5.300}$$

and thus Set 1 becomes

$$\nabla_t^2 E_z + k^2 E_z = i\omega\mu_0 J_z \tag{5.301}$$

$$\mathbf{H}_t = -\frac{1}{i\omega\mu_0}\nabla_t \times \hat{z} E_z = \frac{1}{i\omega\mu_0}\hat{z} \times \nabla_t E_z \tag{5.302}$$

where k is defined in (5.108). The procedure is now to solve the partial differential equation in (5.301) to yield E_z. The result can then be substituted into (5.302) to produce the magnetic fields $\mathbf{H}_t$.

Anticipating the symmetry of the Green's function, we adapt (5.122) to the present case and obtain

$$\int_V g\left(\nabla^2 + k^2\right) E_z dV = \int_V E_z\left(\nabla^2 + k^2\right) g dV + \int_S \left(g\nabla E_z - E_z \nabla g\right)\cdot \hat{n} dS \tag{5.303}$$

The volume V consists of all space exterior to the cylinder. The surface S is the surface of the cylinder S_c plus the surface at infinity. By the same reasoning as in the case of the parallel plate waveguide, the surface integrals at $z \to \pm\infty$ vanish. The Green's function g is governed by

$$-\left(\nabla_t^2 + k^2\right) g = \frac{\delta(\rho - \rho')}{\rho}\delta(\phi - \phi') = \delta(\boldsymbol{\rho} - \boldsymbol{\rho}') \tag{5.304}$$

with boundary and/or limiting conditions to be determined. The conditions on E_z are as follows:

$$E_z|_{S_c} = 0 \tag{5.305}$$

$$\lim_{\rho\to\infty} E_z = 0 \tag{5.306}$$

We choose the following condition for the Green's function g:

$$\lim_{\rho\to\infty} g = 0 \tag{5.307}$$

Therefore, the Green's function problem is for two-dimensional free space. We substitute (5.305)–(5.307) into (5.303) and obtain

$$\int_V g\left(\nabla^2 + k^2\right) E_z dV = \int_V E_z \left(\nabla^2 + k^2\right) g dV + \int_{S_c} g \nabla E_z \cdot \hat{n} dS$$

We note that we did *not* require

$$g|_{S_c} = 0$$

Although such a requirement would eliminate the surface integral, we would be unable to find an analytical solution for the Green's function, except in the special case where the cross section is circular. (We shall consider the circular case subsequently.) Because of the invariance with z, the integrations with respect to z cancel in all terms. In addition, from (5.290), the del-operator reduces to ∇_t and we have

$$\int_A g\left(\nabla_t^2 + k^2\right) E_z dA = \int_A E_z \left(\nabla_t^2 + k^2\right) g dA + \int_{s_c} g \nabla_t E_z \cdot \hat{n} ds \tag{5.308}$$

where s_c is the arc-length integration around the cross section of the cylinder and A is the planar area external to s_c. We shall consider the ds integration in some detail subsequently. Substitution of (5.301) and (5.304) into (5.308) gives, after some rearrangement,

$$E_z(\boldsymbol{\rho}') = -i\omega\mu_0 \int_A g(\boldsymbol{\rho}, \boldsymbol{\rho}') J_z(\boldsymbol{\rho}) dA + \int_{s_c} g(\boldsymbol{\rho}, \boldsymbol{\rho}') \nabla_t E_z(\boldsymbol{\rho}) \cdot n ds \tag{5.309}$$

However, from (5.302), we have

$$\hat{z} \times \nabla_t E_z = i\omega\mu_0 \mathbf{H}_t$$

so that

$$\hat{z} \times (\hat{z} \times \nabla_t E_z) = i\omega\mu_0 \hat{z} \times \mathbf{H}_t$$

But,

$$\hat{z} \times (\hat{z} \times \nabla_t E_z) = -\nabla_t E_z$$

where we have used the vector triple product identity and the fact that

$$\hat{z} \cdot \nabla_t E_z = 0$$

Taking the inner product with the normal vector $\hat{n}$ gives

$$\hat{n} \cdot \nabla_t E_z = i\omega\mu_0 \hat{n} \cdot \mathbf{H}_t \times \hat{z} = i\omega\mu_0 \hat{z} \cdot (\hat{n} \times \mathbf{H}_t) = i\omega\mu_0 J_{sz} \tag{5.310}$$

where J_{sz} is the equivalent surface current in the z-direction in amps/m. Substituting this result into (5.309) and interchanging the primed and unprimed coordinates, we obtain

$$E_z(\rho) = -i\omega\mu_0 \int_A g(\rho, \rho') J_z(\rho') dA' + i\omega\mu_0 \int_{s_c} g(\rho, \rho') J_{sz}(\rho') ds' \tag{5.311}$$

The Green's function problem given in (5.304) and (5.307) has been previously solved in (4.116), followed by the coordinate transformation indicated in (4.205). We have, including 2π from (4.10),

$$g(\rho, \rho') = \frac{1}{4i} H_0^{(2)}\left(k|\rho - \rho'|\right) \tag{5.312}$$

We shall specialize the source J_z to be a line source of strength I_0 amps, located at $\rho' = \rho_0$, viz.

$$J_z = I_0 \delta(\rho' - \rho_0) \tag{5.313}$$

Substituting (5.312) and (5.313) into (5.311), we have

$$E_z(\rho) = -\frac{i\omega\mu_0 I_0}{4i} H_0^{(2)}(k|\rho - \rho_0|) + \frac{i\omega\mu_0}{4i} \int_{s_c} H_0^{(2)}\left(k|\rho - \rho'|\right) J_{sz}(\rho') ds' \tag{5.314}$$

This equation gives the electric field E_z everywhere exterior to the cylinder, provided that we can determine the surface current J_{sz} on the surface of the cylinder. We accomplish this by forming an integral equation. We let ρ approach a general point on the surface of the cylinder $\rho \in s_c$. Since $E_z = 0$ on the cylinder surface, we have

$$I_0 H_0^{(2)}\left(k|\rho - \rho_0|\right) = \int_{s_c} H_0^{(2)}\left(k|\rho - \rho'|\right) J_{sz}(\rho') ds', \qquad \rho \in s_c \tag{5.315}$$

Equations (5.314) and (5.315) complete the formulation of the problem.

Considerable care must be taken in the evaluation of the arc-length integral in (5.314) and (5.315). In Cartesian coordinates, the differential element can be represented by

$$ds' = \left(dx'^2 + dy'^2\right)^{1/2} \tag{5.316}$$

We shall parametrize with respect to the polar angle ϕ' (Fig. 5-10), as follows:

$$ds' = \left[\left(\frac{dx'}{d\phi'}\right)^2 + \left(\frac{dy'}{d\phi'}\right)^2\right]^{1/2} d\phi' \tag{5.317}$$

We locate the origin of the coordinate system (Fig. 5-10) internal to s_c. Substituting (5.317) into (5.314) and (5.315), we produce

$$E_z(\boldsymbol{\rho}) = -\frac{\omega\mu_0 I_0}{4} H_0^{(2)}\left(k|\boldsymbol{\rho} - \boldsymbol{\rho}_0|\right) + \frac{\omega\mu_0}{4}\int_0^{2\pi} H_0^{(2)}(k|\boldsymbol{\rho} - \boldsymbol{\rho}'|)\, J_{sz}(\boldsymbol{\rho}')\left[\left(\frac{dx'}{d\phi'}\right)^2 + \left(\frac{dy'}{d\phi'}\right)^2\right]^{1/2} d\phi' \tag{5.318}$$

and for $\boldsymbol{\rho} \in s_c$,

$$I_0 H_0^{(2)}\left(k|\boldsymbol{\rho} - \boldsymbol{\rho}_0|\right) = \int_0^{2\pi} H_0^{(2)}\left(k|\boldsymbol{\rho} - \boldsymbol{\rho}'|\right) J_{sz}(\boldsymbol{\rho}') \cdot \left[\left(\frac{dx'}{d\phi'}\right)^2 + \left(\frac{dy'}{d\phi'}\right)^2\right]^{1/2} d\phi' \tag{5.319}$$

As in the previous problems in this chapter, we must solve the integral equation to determine the unknown quantity under the integral, in this case, J_{sz}. In general, numerical methods must be employed to obtain an approximation to the solution. In the case where the cylindrical cross section is circular, however, we may invert the integral equation in (5.319) analytically. Indeed, consider a perfectly conducting circular cylinder of radius a (Fig. 5-11). In this case,

$$x' = \rho' \cos\phi'$$
$$y' = \rho' \sin\phi'$$
$$\frac{dx'}{d\phi'} = -\rho' \sin\phi'$$
$$\frac{dy'}{d\phi'} = \rho' \cos\phi'$$

Substitution in (5.317) produces the usual cylindrical representation of the arc-length integration

$$ds' = \rho' d\phi' \tag{5.320}$$

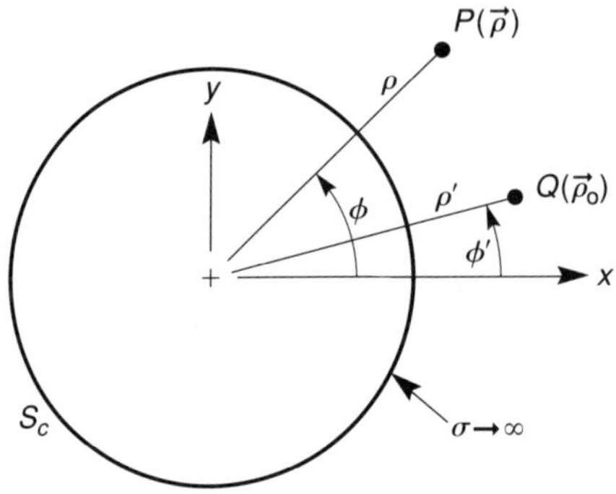

Fig. 5-11 Electric current J_z exciting a perfectly conducting circular cylinder.

We shall employ the addition theorem for the Hankel function from (4.207), viz.

$$H_0^{(2)}\left(k|\boldsymbol{\rho}-\boldsymbol{\rho}'|\right)=\sum_{n=-\infty}^{\infty} e^{in(\phi-\phi')}\begin{cases} H_n^{(2)}(k\rho')J_n(k\rho), & \rho<\rho' \\ H_n^{(2)}(k\rho)J_n(k\rho'), & \rho>\rho' \end{cases} \tag{5.321}$$

Substituting (5.320) and (5.321) into (5.319) and rearranging, we have

$$\sum_{n=-\infty}^{\infty}\left[I_0 e^{-in\phi_0}H_n^{(2)}(k\rho_0)J_n(ka)\right]e^{in\phi}$$
$$=\sum_{n=-\infty}^{\infty}\left[\int_0^{2\pi} J_{sz}(a,\phi')e^{-in\phi'}a\,d\phi' H_n^{(2)}(ka)J_n(ka)\right]e^{in\phi} \tag{5.322}$$

To obtain (5.322), since $\boldsymbol{\rho}\in s_c$, we have chosen the $\rho<\rho'$ case in (5.321). We recognize each side of (5.322) as a complex Fourier series on $(0, 2\pi)$. We equate coefficients and rearrange to give

$$\int_0^{2\pi} J_{sz}(a,\phi')e^{-in\phi'}d\phi'=\frac{I_0 e^{-in\phi_0}H_n^{(2)}(k\rho_0)}{aH_n^{(2)}(ka)} \tag{5.323}$$

We shall use this result to find the electric field E_z by noting that, in (5.318),

$$\int_0^{2\pi} H_0^{(2)}\left(k|\boldsymbol{\rho}-\boldsymbol{\rho}'|\right) J_{sz}(\boldsymbol{\rho}')a\,d\phi'$$
$$=\sum_{n=-\infty}^{\infty} e^{in\phi}H_n^{(2)}(k\rho)J_n(ka)\int_0^{2\pi} J_{sz}(a,\phi')e^{-in\phi'}a\,d\phi'$$
$$=I_0\sum_{n=-\infty}^{\infty} e^{in(\phi-\phi_0)}\frac{J_n(ka)}{H_n^{(2)}(ka)}H_n^{(2)}(k\rho)H_n^{(2)}(k\rho_0) \tag{5.324}$$

Substituting this result and (5.320) into (5.314), we obtain the following expansion for the electric field E_z:

$$E_z(\boldsymbol{\rho}) = -\frac{\omega\mu_0 I_0}{4}\left\{H_0^{(2)}\left(k|\boldsymbol{\rho}-\boldsymbol{\rho}_0|\right) - \sum_{n=-\infty}^{\infty} e^{in(\phi-\phi_0)} \cdot \frac{J_n(ka)}{H_n^{(2)}(ka)} H_n^{(2)}(k\rho) H_n^{(2)}(k\rho_0)\right\} \tag{5.325}$$

which is the classical ϕ-directed eigenfunction expansion for the electric field [15].

In the above example of the circular cylinder, we were able to invert the integral equation analytically. This event occurred because the surface S_c was a coordinate surface, in this case, $\rho = a$. In cases where the cylinder does not conform to a complete coordinate surface, the integral equation in (5.315) must be inverted numerically. We include a specific case, the rectangular cylinder, in the problems.

5.9 PERFECTLY CONDUCTING CIRCULAR CYLINDER

In the previous section, we derived the fields associated with scattering from a perfectly conducting circular cylinder by beginning with the conducting cylinder of arbitrary cross section. We obtained an integral equation in (5.319). For the case of circular cross section, we were able to invert the integral equation and obtain an expression for the electric field E_z in (5.325). It is, however, possible to proceed more directly. In this section, we derive the fields scattered from a perfectly conducting cylinder of circular cross section when the excitation is an electric current line source. We are able to verify the result obtained in (5.325). Next, we obtain an alternative representation, useful in describing scattering in the form of *creeping waves.*

We again consider the geometry in Fig. 5-11. The source is given explicitly in (4.163) and the differential equation describing the E_z-field in (4.180) and (4.181), which we repeat here for convenience, viz.

$$\frac{1}{\rho}\left[\frac{\partial}{\partial\rho}\left(\rho\frac{\partial g}{\partial\rho}\right)\right] + \frac{1}{\rho^2}\frac{\partial^2 g}{\partial\phi^2} + k^2 g = -\frac{\delta(\rho-\rho')\delta(\phi-\phi')}{\rho} \tag{5.326}$$

$$g = -\frac{E_z}{i\omega\mu_0 I_0} \tag{5.327}$$

From the results in Problem 3.2, we may expand the Green's function g in terms of the spectral representation with respect to ϕ, viz.

$$g(\rho, \phi, \rho', \phi') = \sum_{n=-\infty}^{\infty} a_n(\rho, \rho', \phi') \sqrt{\frac{1}{2\pi}} e^{in\phi} \tag{5.328}$$

We write this transformation

$$g \Longleftrightarrow a_n \tag{5.329}$$

and easily find that

$$-\frac{\partial^2 g}{\partial \phi^2} \Longleftrightarrow n^2 a_n \tag{5.330}$$

$$\delta(\phi - \phi') \Longleftrightarrow \sqrt{\frac{1}{2\pi}} e^{-in\phi'} \tag{5.331}$$

Applying (5.329)–(5.331) to (5.326), we obtain

$$\frac{1}{\rho}\left[\frac{d}{d\rho}\left(\rho \frac{db_n}{d\rho}\right)\right] + k^2 b_n - \frac{n^2}{\rho^2} b_n = -\frac{\delta(\rho - \rho')}{\rho} \tag{5.332}$$

where

$$a_n = \sqrt{\frac{1}{2\pi}} e^{-in\phi'} b_n \tag{5.333}$$

So far, the development is identical to that in (4.190)–(4.198), except that the boundary and limiting conditions are now

$$b_n \Big|_{\rho=a} = 0 \tag{5.334}$$

$$\lim_{\rho \to \infty} b_n = 0 \tag{5.335}$$

We write the solution for b_n as a linear combination of Bessel and Hankel functions, as follows:

$$b_n = \begin{cases} A J_n(k\rho) + C H_n^{(2)}(k\rho), & \rho < \rho' \\ B H_n^{(2)}(k\rho) + D H_n^{(1)}(k\rho), & \rho > \rho' \end{cases} \tag{5.336}$$

The limiting condition in (5.335) results in $D = 0$. At $\rho = a$, we have

$$A J_n(ka) + C H_n^{(2)}(ka) = 0$$

Solving for C and substituting into (5.336), we have

$$b_n = \begin{cases} A\left[J_n(k\rho) - c_n H_n^{(2)}(k\rho)\right], & \rho < \rho' \\ BH_n^{(2)}(k\rho), & \rho > \rho' \end{cases} \tag{5.337}$$

where

$$c_n = \frac{J_n(ka)}{H_n^{(2)}(ka)} \tag{5.338}$$

Invoking the continuity and jump conditions at $\rho = \rho'$ allows us to evaluate the coefficients A and B, as follows:

$$A = \frac{\pi}{2i} H_n^{(2)}(k\rho') \tag{5.339}$$

$$B = \frac{\pi}{2i}\left[J_n(k\rho') - c_n H_n^{(2)}(k\rho')\right] \tag{5.340}$$

where we have used the Bessel function identity

$$J_n H_n^{(2)\prime} - J_n{}' H_n^{(2)} = \frac{2}{i\pi x} \tag{5.341}$$

where the prime indicates differentiation with respect to x. Therefore,

$$b_n = \frac{\pi}{2i} \begin{cases} H_n^{(2)}(k\rho')\left[J_n(k\rho) - c_n H_n^{(2)}(k\rho)\right], & \rho < \rho' \\ H_n^{(2)}(k\rho)\left[J_n(k\rho') - c_n H_n^{(2)}(k\rho')\right], & \rho > \rho' \end{cases} \tag{5.342}$$

Substitution of (5.342) into (5.333) and the result into (5.328) gives

$$g = \frac{1}{4i} \sum_{n=-\infty}^{\infty} e^{in(\phi-\phi')} \left[-c_n H_n^{(2)}(k\rho) H_n^{(2)}(k\rho') + \begin{cases} H_n^{(2)}(k\rho') J_n(k\rho), & \rho < \rho' \\ H_n^{(2)}(k\rho) J_n(k\rho'), & \rho > \rho' \end{cases} \right] \tag{5.343}$$

Use of (5.327) and the addition theorem given in (4.207) again produces the result in (5.325), which we display here for reference, viz.

$$E_z(\boldsymbol{\rho}) = -\frac{\omega\mu_0 I_0}{4}\left\{ H_0^{(2)}\left(k|\boldsymbol{\rho} - \boldsymbol{\rho}_0|\right) - \sum_{n=-\infty}^{\infty} e^{in(\phi-\phi_0)} \cdot \frac{J_n(ka)}{H_n^{(2)}(ka)} H_n^{(2)}(k\rho) H_n^{(2)}(k\rho_0) \right\} \tag{5.344}$$

It is instructive to consider the important special case of a plane wave incident on the cylinder. In (5.344), the first term in the brackets is the incident field, given by

$$E_z^{inc} = -\frac{\omega_0 \mu_0 I_0}{4} H_0^{(2)}(k|\boldsymbol{\rho} - \boldsymbol{\rho}'|) \tag{5.345}$$

We expand $|\boldsymbol{\rho} - \boldsymbol{\rho}'|$ in cylindrical coordinates and obtain

$$|\boldsymbol{\rho} - \boldsymbol{\rho}'| = \left[\rho^2 + \rho'^2 - 2\rho\rho' \cos(\phi - \phi')\right]^{1/2} \tag{5.346}$$

The plane wave case is produced by allowing the line source to be very far removed from the cylinder. Mathematically, $\rho' >> \rho$ and

$$\begin{aligned} |\boldsymbol{\rho} - \boldsymbol{\rho}'| &\cong \rho' \left[1 - 2\left(\frac{\rho}{\rho'}\right) \cos(\phi - \phi')\right]^{1/2} \\ &\cong \rho' - \rho \cos(\phi - \phi') \end{aligned} \tag{5.347}$$

where we have discarded terms in ρ/ρ' higher than first order, and where we have used the first two terms in the Taylor series expansion for $\sqrt{(1+x)}$. Substituting (5.347) into (5.345), and using the large argument approximation for the Hankel function given in Example 2.21, we obtain

$$E_z^{inc} = -\frac{i\omega\mu_0 I_0}{4i} \sqrt{\frac{2i}{\pi k \rho'}} e^{-ik[\rho' - \rho \cos(\phi - \phi')]} \tag{5.348}$$

We let the incident wave arrive from left to right along the x-axis (Fig. 5-11), so that $\phi' = \pi$ and

$$E_z^{inc} = -\frac{i\omega\mu_0 I_0}{4i} \sqrt{\frac{2i}{\pi k \rho'}} e^{-ik\rho'} e^{-ik\rho \cos\phi} \tag{5.349}$$

To produce a unit magnitude plane wave from left to right, we adjust the intensity I_0 as follows:

$$I_0 = \left[-\frac{i\omega\mu_0 I_0}{4i} \sqrt{\frac{2i}{\pi k \rho'}} e^{-ik\rho'}\right]^{-1} \tag{5.350}$$

so that

$$E_z^{inc} = e^{-ik\rho \cos\phi} = e^{-ikz} \tag{5.351}$$

In obtaining the fields for TM propagation between parallel plates in Section 5.5, we found that there was an alternative representation for the Green's function, useful at high frequencies. In the case under consideration here, we may again obtain a useful alternative representation. We begin by writing the differential equation describing the Green's function in (5.326) in the form that separates the ρ-operator from the ϕ-operator. This form is given in (4.187)–(4.189) and is repeated here for convenience, viz.

$$(L_\rho + L_\phi)g = \rho\delta(\rho - \rho')\delta(\phi - \phi') \tag{5.352}$$

where

$$L_\rho = -\rho\left[\frac{\partial}{\partial\rho}\left(\rho\frac{\partial}{\partial\rho}\right)\right] - (k\rho)^2 \tag{5.353}$$

$$L_\phi = -\frac{\partial^2}{\partial\phi^2} \tag{5.354}$$

We require the spectral representation of the operator L_ρ. The Green's function problem associated with this spectral representation is

$$(L_\rho - \lambda)G = \rho\delta(\rho - \rho') \tag{5.355}$$

$$G\Big|_{\rho=a} = 0 \tag{5.356}$$

$$\lim_{\rho\to\infty} G = 0 \tag{5.357}$$

The reader should carefully compare this problem to the problem in Example 3.6. The only difference is in the boundary condition at the lower end of the interval. In Example 3.6, we had a finiteness condition at $\rho = 0$. In this case, we have a Dirichlet condition at $\rho = a$. We still have the limit point case as $\rho \to \infty$, but the condition at $\rho = a$ is regular. We write the solution as

$$G = \begin{cases} AJ_\nu(k\rho) + CH_\nu^{(2)}(k\rho), & \rho < \rho' \\ BH_\nu^{(2)}(k\rho) + DH_\nu^{(1)}(k\rho), & \rho > \rho' \end{cases} \tag{5.358}$$

where

$$\nu = i\sqrt{\lambda} \tag{5.359}$$

in the same manner as in (3.136). Application of the limiting condition in (5.357) results in $D = 0$. From this point, the solution for the Green's function follows the development in (5.337)–(5.342). The result is

$$G = \frac{\pi}{2i} \begin{cases} H_\nu^{(2)}(k\rho') \left[J_\nu(k\rho) - c_\nu H_\nu^{(2)}(k\rho) \right], & \rho < \rho' \\ H_\nu^{(2)}(k\rho) \left[J_\nu(k\rho') - c_\nu H_\nu^{(2)}(k\rho') \right], & \rho > \rho' \end{cases} \tag{5.360}$$

where the branch cut in $\sqrt{\lambda}$ lies along the positive real axis and is explicitly determined by (3.143).

Our next step is to determine the spectral representation of $\rho\delta(\rho - \rho')$ by integrating the Green's function with respect to λ by the methods developed in Chapter 3. We first consider the case $\rho < \rho'$. We have

$$G = \frac{\pi}{2i} H_{i\sqrt{\lambda}}^{(2)}(k\rho') \left[J_{i\sqrt{\lambda}}(k\rho) - \frac{J_{i\sqrt{\lambda}}(ka)}{H_{i\sqrt{\lambda}}^{(2)}(ka)} H_{i\sqrt{\lambda}}^{(2)}(k\rho) \right] \tag{5.361}$$

where we have used (5.338) and (5.359). We define the branch cut associated with $\sqrt{\lambda}$ by using (3.143) and (3.144), and produce a cut along the positive-real axis in the λ-plane. Following the development in Example 3.6, we now investigate whether the branch cut in $\sqrt{\lambda}$ produces a branch cut in G. As we approach the positive- real axis from above and below, we have, respectively,

$$\lim_{\phi \to -2\pi} G = \frac{\pi}{2i} \frac{H_{-\tau}^{(2)}(k\rho')}{H_{-\tau}^{(2)}(ka)} \left[J_{-\tau}(k\rho) H_{-\tau}^{(2)}(ka) - J_{-\tau}(ka) H_{-\tau}^{(2)}(k\rho) \right] \tag{5.362}$$

$$\lim_{\phi \to 0} G = \frac{\pi}{2i} \frac{H_{\tau}^{(2)}(k\rho')}{H_{\tau}^{(2)}(ka)} \left[J_{\tau}(k\rho) H_{\tau}^{(2)}(ka) - J_{\tau}(ka) H_{\tau}^{(2)}(k\rho) \right] \tag{5.363}$$

where

$$\tau = i|\lambda|^{1/2} \tag{5.364}$$

But, from [16], we have

$$J_\tau = \frac{1}{2} \left[H_\tau^{(1)} + H_\tau^{(2)} \right] \tag{5.365}$$

$$J_{-\tau} = \frac{1}{2} \left[e^{i\pi\tau} H_\tau^{(1)} + e^{-i\pi\tau} H_\tau^{(2)} \right] \tag{5.366}$$

and, from (3.149),

$$H_{-\tau}^{(2)} = e^{-i\pi\tau} H_\tau^{(2)} \tag{5.367}$$

We substitute (5.365) into (5.363); in addition, we substitute (5.366) and (5.367) into (5.362). After some routine algebra, we find that

$$\lim_{\phi \to -2\pi} G = \lim_{\phi \to 0} G \tag{5.368}$$

Therefore, there is no branch cut in G along the positive- real axis. Since G has no branch cut singularities, the spectral representation of the delta function is given by (3.39), viz.

$$-\rho\delta(\rho - \rho') = \frac{1}{2\pi i} \oint G(\rho, \rho', \lambda) d\lambda \tag{5.369}$$

where the only possible singularities in G are poles. Our analysis of the pole contributions is based on the treatments in [17] and [18]. We write the expression for G given in (5.361) as

$$G = G_1 + G_2 \tag{5.370}$$

where

$$G_1 = \frac{\pi}{2i} H^{(2)}_{i\sqrt{\lambda}}(k\rho') J_{i\sqrt{\lambda}}(k\rho) \tag{5.371}$$

$$G_2 = -\frac{\pi}{2i} \frac{J_{i\sqrt{\lambda}}(ka)}{H^{(2)}_{i\sqrt{\lambda}}(ka)} H^{(2)}_{i\sqrt{\lambda}}(k\rho) H^{(2)}_{i\sqrt{\lambda}}(k\rho') \tag{5.372}$$

There are no poles contained in G_1; there are, however, a countably infinite number of simple poles [18] in G_2 whenever

$$H^{(2)}_{i\sqrt{\lambda}}(ka) = 0 \tag{5.373}$$

Therefore, by Cauchy's Theorem, only the second term in (5.361) contributes to the contour integral in (5.369), and we have

$$\rho\delta(\rho - \rho') = -\frac{1}{4} \oint \frac{J_{i\sqrt{\lambda}}(ka)}{H^{(2)}_{i\sqrt{\lambda}}(ka)} H^{(2)}_{i\sqrt{\lambda}}(k\rho) H^{(2)}_{i\sqrt{\lambda}}(k\rho') d\lambda \tag{5.374}$$

Using the residue theorem, we obtain

$$\rho\delta(\rho - \rho') = \frac{\pi}{2i} \sum_{p=1}^{\infty} J_{i\sqrt{\lambda_p}}(ka) H^{(2)}_{i\sqrt{\lambda_p}}(k\rho) H^{(2)}_{i\sqrt{\lambda_p}}(k\rho') \mathrm{Res}\left[\frac{1}{H^{(2)}_{i\sqrt{\lambda_p}}}; \lambda_p \right] \tag{5.375}$$

where

$$\mathrm{Res}[f(z); z]$$

signifies the residue of $f(z)$ evaluated at z, and where the sum is over the zeros evaluated in (5.373). Using the relationship between λ and ν in (5.359), we obtain

$$\rho\delta(\rho - \rho') = \sum_{p=1}^{\infty} \Phi_p(k\rho)\Phi_p(k\rho') \tag{5.376}$$

where

$$\Phi_p(\rho) = \left\{ \pi i \frac{\nu_p J_{\nu_p}(ka)}{\frac{\partial}{\partial\nu}\left[H_\nu^{(2)}(ka)\right]_{\nu_p}} \right\}^{1/2} H_{\nu_p}^{(2)}(k\rho) \tag{5.377}$$

Expression (5.376) gives the required spectral representation for the delta function. We note that the result is symmetric with respect to ρ and ρ'. Therefore, the restriction $\rho < \rho'$ can be removed.

Using the methods in Chapter 3, we may develop the spectral representation in (5.376) into a Fourier expansion useful for solving (5.352). For $f(\rho) \in \mathcal{L}_2(a, \infty)$, we have

$$f(\rho) = \int_a^\infty f(\rho')\delta(\rho-\rho')d\rho' = \int_a^\infty f(\rho')\left[\rho'\delta(\rho - \rho')\right]\frac{d\rho'}{\rho'} \tag{5.378}$$

Since $\rho'\delta(\rho - \rho') = \rho\delta(\rho - \rho')$, we may substitute (5.376) and obtain

$$f(\rho) = \sum_{p=1}^{\infty} \alpha_p \Phi_p(k\rho) \tag{5.379}$$

where

$$\alpha_p = \int_a^\infty f(\rho)\Phi_p(k\rho)\frac{d\rho}{\rho} \tag{5.380}$$

We next use this Fourier expansion to solve (5.352). Let

$$g = \sum_{p=1}^{\infty} \alpha_p(\rho', \phi, \phi')\Phi(k\rho) \tag{5.381}$$

Substitution into (5.352) gives

$$\left(\frac{d^2}{d\phi^2} + \nu_p^2\right)\alpha_p = -\Phi_p(k\rho')\delta(\phi - \phi') \tag{5.382}$$

This Green's function problem with periodic boundary conditions has been solved in Problem 2.18. The result applied here is

$$\alpha_p = -\frac{\cos[\nu_p(|\phi - \phi'| - \pi)]}{2\nu_p \sin \nu_p\pi}\Phi_p(k\rho') \tag{5.383}$$

We substitute into (5.381) and find that

$$g = -\sum_{p=1}^{\infty} \frac{\cos[\nu_p(|\phi - \phi'| - \pi)]}{2\nu_p \sin \nu_p \pi} \Phi_p(k\rho)\Phi_p(k\rho') \tag{5.384}$$

Using (5.327), we produce the electric field

$$E_z = i\omega\mu_0 I_0 \sum_{p=1}^{\infty} \frac{\cos[\nu_p(|\phi - \phi'| - \pi)]}{2\nu_p \sin \nu_p \pi} \Phi_p(k\rho)\Phi_p(k\rho') \tag{5.385}$$

We again specialize to the case where a plane wave is incident from left to right along the x-axis (Fig. 5-11). Let

$$\phi' = \pi, \qquad -\pi < \phi \le \pi \tag{5.386}$$

so that

$$\cos[\nu_p(|\phi - \phi'| - \pi)] = \cos \nu_p \phi \tag{5.387}$$

Let ρ' become large enough so that the Hankel function can be approximated by

$$H_{\nu_p}^{(2)}(k\rho') \sim \sqrt{\frac{2i}{\pi k\rho'}} i^{\nu_p} e^{-ik\rho'} \tag{5.388}$$

Then,

$$E_z = -\frac{\omega\mu_0 \pi I_0}{2} \sqrt{\frac{2i}{\pi k\rho'}} e^{-ik\rho'} \sum_{p=1}^{\infty} i^{\nu_p} \frac{J_{\nu_p}(ka) H_{\nu_p}^{(2)}(k\rho) \cos \nu_p \phi}{\frac{\partial}{\partial \nu}[H_\nu^{(2)}(ka)]_{\nu_p} \sin \nu_p \pi} \tag{5.389}$$

To produce a unit plane wave incident, we use (5.350) and produce

$$E_z(\rho, \phi) = 2\pi \sum_{p=1}^{\infty} i^{\nu_p} \frac{J_{\nu_p}(ka) H_{\nu_p}^{(2)}(k\rho) \cos \nu_p \phi}{\frac{\partial}{\partial \nu}[H_\nu^{(2)}(ka)]_{\nu_p} \sin \nu_p \pi} \tag{5.390}$$

The reader may wish to compare this result with [17, eq. (129)] by using (5.365) and (5.373). In [17], James has used the classic residue series approach to produce the representation for the electric field that we give in (5.390). We have used the alternative spectral representation, a method also used in [18].

The alternative spectral representation is useful for obtaining solutions at high frequencies where summing the series in (5.344) requires a large number of terms for convergence. The alternative representation is particularly suited for the so-called *shadow region* [19] behind the cylinder, away from the side directly illuminated by the incoming plane wave. Here, the field is given in the form of *creeping waves* [17], [20]. For a thorough discussion of the zeros of the Hankel function needed in (5.373), the reader is referred to [18] and [21].

5.10 DYADIC GREEN'S FUNCTIONS

In the electromagnetic problems in this chapter, the geometry and source in each case have been independent of one coordinate dimension. These two-dimensional problems have been chosen as models to illustrate the use of spectral expansions and Green's functions. Indeed, many of the interesting and useful problems in electromagnetic theory can be modeled in two spatial dimensions. Additional two-dimensional examples directly using the methods developed in this book can be found in [7]–[9].

There are, however, many electromagnetic problems where it is not feasible to assume that the problem is independent of one spatial dimension. In these three-dimensional cases, the analysis in this book may be directly and elegantly extended using a dyadic form of Green's theorem. The dyadic method is presented in detail in the book by Tai [22].

Dyadic analysis is based on the formulation of dyadic spectral representations of the delta function and the derivation of problem-dependent dyadic Green's functions. The interested reader is referred to [22] for a description of the procedures, as well as application to some classical problems, such as waveguide propagation, scattering from cylinders, and interactions with plane stratified media. In addition, the book by Collin [23] provides a logical, systematic presentation of dyadic Green's functions and their use in electromagnetics.

Dyadic analysis can be applied to boundary value problems where the solution depends on inverting an integral equation. In these cases, the reader is cautioned that the analysis of the singularities associated with dyadic kernels in integral equations is a delicate matter. For a discussion, the reader is referred to [23],[24].

PROBLEMS

5.1. Using the Green's function method, show that (5.51) is 6the solution to (5.49) with the boundary conditions in (5.50).

5.2. Show that the solution in (5.72) satisfies the differential equation in (5.55).

5.3. Beginning with the TE_z equation set in (5.109)–(5.111), derive a modal series dual to the modal series describing the TM_z modes in (5.138).

5.4. Using the Green's function method, show that (5.199) is the solution to (5.196) with the boundary conditions in (5.197) and (5.198).

5.5. In the problem describing the scattering from a perfectly conducting cylinder given in Section 5.8, assume that the cylinder cross section is rectangular. For this specific case, specialize the expression for the electric field in (5.318) and the form of the integral equation in (5.319).

5.6. Consider a y-directed magnetic current source M_y above an impedance plane (Fig. 5-12). Let the current source be independent of y, so that

$$\frac{\partial}{\partial y} = 0$$

Assume that the boundary condition at the impedance plane is given by

$$\lim_{x \to 0} \left(\frac{E_z}{H_y} \right) = i\omega L$$

where $L > 0$ is the inductance of the impedance sheet.

(a) Show that the only nonzero field components are H_y, E_x, E_z.

(b) Using Green's theorem, formulate an expression for the magnetic field H_y. Solve explicitly for the Green's function by two methods:

1. Use a spectral expansion in z, followed by a closed form solution in x.
2. Use a spectral expansion in x, followed by a closed form solution in z. *Note*: This spectral expansion utilizes the impedance transform derived in Chapter 3.

(c) Specialize the solution to the case where the magnetic current source is a line source on the x-axis at a distance d above the impedance plane.

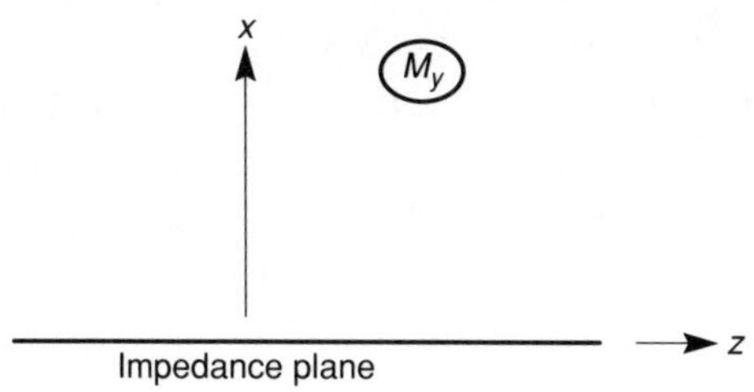

Fig. 5-12 Magnetic current M_y above an impedance plane.

REFERENCES

[1] Dennemeyer, R. (1968), *Introduction to Partial Differential Equations*. New York: McGraw-Hill, 92–93.

[2] Stratton, J.A. (1941), *Electromagnetic Theory*. New York: McGraw-Hill, 165.

[3] Ramo, S., J.R. Whinnery, and T. Van Duzer (1984), *Fields and Waves in Communication Electronics*. New York: Wiley, 396–399.

[4] Cheng, D.K. (1989), *Field and Wave Electromagnetics*. Reading, MA: Addison-Wesley, 534–539.

[5] Felsen, L.B., and A.H. Kamel (1981), Hybrid ray-mode formulation of parallel plate waveguide Green's functions, *IEEE Trans. Antennas Propagat.* AP-29: 637–649.

[6] Wilton, D.R., and C.M. Butler (1990), Effective methods for solving integral and integrodifferential equations, *Electromagnetics* 1: 289–308; reprinted in Hansen, R.C. (Ed.), *Moment Methods in Antennas and Scattering*. Norwood, MA: Artech House, 58–77.

[7] Zhang, S.J., and D.G. Dudley (1987), Scattering by a radial line joined to a coaxial waveguide, *J. Electromagnetic Waves and Appl.* 1: 299–321.

[8] Lee, R., D.G. Dudley, and K.F. Casey (1989), Electromagnetic coupling by a wire through a circular aperture in an infinite planar screen, *J. Electromagnetic Waves and Appl.* 3: 281–305.

[9] Wright, D.B., D.G. Dudley, and K.F. Casey (1990), Electromagnetic coupling by a wire through a cavity-backed circular aperture in an infinite screen, *J. Electromagnetic Waves and Appl.* 4: 375–400.

[10] Collin, R.E. (1991), *Field Theory of Guided Waves*, 2nd edition. New York: IEEE Press, 811–821.

[11] Mittra, R., and S.W. Lee (1971), *Analytical Techniques in the Theory of Guided Waves*. New York: Macmillan, sect. 1-3.

[12] Butler, C.M. (1984), Current induced on a conducting strip which resides on the planar interface between two semi-infinite half-spaces, *IEEE Trans. Antennas Propagat.* AP-32: 226–231.

[13] Dudley, D.G. (1985), Error minimization and convergence in numerical methods, *Electromagnetics* 5: 89–97.

[14] Butler, C.M., and D.R. Wilton (1980), General analysis of narrow strips and slots, *IEEE Trans. Antennas Propagat.* AP-28: 42–48.

[15] Harrington, R.F. (1961), *Time-Harmonic Electromagnetic Fields.* New York: McGraw-Hill, 232–238.

[16] Abramowitz, M., and I.A. Stegun (Eds.) (1964), *Handbook of Mathematical Functions*, National Bureau of Standards, Applied Mathematics Series, 55, Superintendent of Documents, U.S. Government Printing Office, Washington, DC 20402, 358, eq. 9.1.3 and 9.1.4.

[17] James, G.L. (1980), *Geometric Theory of Diffraction for Electromagnetic Waves.* Stevenage, UK: Peter Peregrinus, 74–76.

[18] Felsen, L.B., and N. Marcuvitz (1973), *Radiation and Scattering of Waves.* Englewood Cliffs, NJ: Prentice-Hall, 326–327, 710–718.

[19] Ishimaru, A. (1991), *Electromagnetic Wave Propagation, Radiation, and Scattering.* Englewood Cliffs: NJ, Prentice-Hall, 330–333.

[20] Franz, W. (1954), Über die greensche funktionen des zylinders und der kugel, *Z. Naturforsch* 9: 705–716.

[21] Wait, J.R. (1988), *Electromagnetic Radiation from Cylindrical Structures* Stevenage, UK: Peter Peregrinus, 54–60.

[22] Tai, C.T. (1993), *Dyadic Green's Functions in Electromagnetic Theory*, 2nd edition. New York: IEEE Press.

[23] op.cit. Collin, chapter 2.

[24] Johnson, W.A., A.Q. Howard, and D.G. Dudley (1979), On the irrotational component of the electric Green's dyadic, *Radio Science* 14, 961–967.

Index

F

G

H

I

R

S